The SpecOps Method:
A New Approach to Modernizing
Legacy Technology Systems

Mark Headd

Civic Innovations Press

The SpecOps Method: A New Approach to Modernizing Legacy Technology Systems

Published by Civic Innovations Press

ISBN: 979-8-9944547-1-8

Online Resources

Website: https://spec-ops.ai/
GitHub: https://github.com/spec-ops-method
Discussion: https://github.com/spec-ops-method/spec-ops/discussions

Credits

Copy Editor: Claude Opus 4.5 (Anthropic)
Cover: Image of a mainframe computer system being erased generated with Gemini 3 (Google)

Disclaimer

The views and opinions expressed in this book are those of the author and do not necessarily reflect the official policy or position of the author's current or former employers.

The information in this book is provided for informational purposes only. The author and publisher make no representations or warranties regarding the accuracy or completeness of the contents and specifically disclaim any implied warranties. The advice and strategies contained herein may not be suitable for your situation. Neither the publisher nor the author shall be liable for any loss of profit or any other commercial damages, including but not limited to special, incidental, consequential, or other damages.

First Edition

To everyone working in service of the public good.

Table of Contents

Introduction

Some of the most critical systems in government run on software that is older than most of the people trying to maintain them.

The systems processing unemployment claims, tax returns, and benefits eligibility were often built decades ago, in programming languages that most universities no longer teach, by people who have retired, or those getting ready to. These systems work, after a fashion. They process millions of transactions. They deliver services that many citizens depend on. But they are fragile, poorly understood, largely undocumented, and increasingly dangerous to change.

This is not a new problem. Government technology leaders and people working in civic tech have known about the problems presented by aging systems for years. What's new is the convergence of two forces that together create both heightened urgency and new opportunity.

The urgency is a function of demographics. Some of the most critical systems operated by federal and state governments are built on technologies that stretch back to the 1950s. The average age of the people in government who have expertise in these antiquated technologies has now crossed the retirement threshold. The people who understand how government systems actually work are leaving at an accelerated rate, and in many cases their knowledge is leaving with them. The window for capturing institutional knowledge about how critical government systems work, and why they work the way they do, is closing.

The opportunity is a result of recent advances in technology. AI coding assistants have reached a threshold of capability where they can analyze legacy code at scale, extract business logic, and generate documentation that humans can review and verify. This doesn't mean AI can magically fix legacy systems. It can't. But it does mean that tasks which were previously impossible, like comprehensively documenting vast legacy codebases, are now merely difficult.

This book introduces a new approach to legacy system modernization called SpecOps that harnesses the opportunity presented by these new tools to address the growing urgency. The basic premise is simple: the valuable aspect of a software system isn't the code it is written in. It's the detailed knowledge of what the system does and why. Code is just one expression of that knowledge, written in a language that becomes obsolete over time. Specifications capture the same knowledge in a form that can be persisted.

SpecOps treats specifications as the source of truth for system behavior. AI assists in extracting specifications from legacy code. Domain experts verify that those specifications correctly describe how systems work. Modern implementations can then be generated from verified specifications, often using AI-coding assistants and similar tools. The result is a modernization approach that preserves institutional knowledge rather than losing it, and that produces systems designed to evolve over time rather than become archaic.

The book is organized into six parts.

Part I: The Legacy Modernization Crisis establishes the problem. Chapter 1 describes the scale of government legacy systems and the workforce dynamics that make modernization urgent. Chapter 2 traces how we got here: the history of government IT, the budget structures that prevent continuous improvement, and the organizational dynamics that accumulate technical debt. Chapter 3 examines the traditional modernization playbook and explains why big-bang replacements, waterfall development, and direct code translation-whether by teams of humans or through the use of AI-have consistently failed.

Part II: The AI-Assisted Development Revolution introduces the enabling technology. Chapter 4 surveys the emergence of AI coding assistants and provides an honest assessment of what they can and can't do. Chapter 5 explores specification-driven development and how AI makes comprehensive specifications practical for the first time. Chapter 6 draws a parallel to infrastructure as code, showing how the discipline of GitOps established the pattern of treating version-controlled declarations as the source of truth for technology implementations.

Part III: Introducing SpecOps presents the SpecOps methodology itself. Chapter 7 distinguishes compilation (preserving knowledge in specifications) from transpilation (translating old code to new code without preserving understanding) and argues that the specification is the valuable artifact. Chapter 8 walks through the six phases of the SpecOps methodology, from Discovery through Deployment. Chapter 9 articulates the core principles that guide the approach and introduces the tools that support it.

Part IV: Why SpecOps Works for Government addresses the specific context of government technology. Chapter 10 examines the knowledge preservation imperative and how specifications serve as institutional memory. Chapter 11 confronts the politics of modernization: the accountability requirements, the risk aversion, and how to build confidence through incremental delivery. Chapter 12 explores collaboration opportunities, showing how instruction sets can be shared across agencies even when code can't be.

Part V: SpecOps in Practice provides practical guidance for teams adopting the SpecOps approach. Chapter 13 covers getting started: selecting a pilot, building a team, and setting up the SpecOps toolchain. Chapter 14 addresses working effectively with both AI tools and human collaborators. Chapter 15 tackles common challenges: what to do when domain experts aren't available, how to handle truly incomprehensible software code, and how to balance speed and thoroughness.

Part VI: Looking Forward considers what success would mean and what comes next. Chapter 16 asks what it would take to break the legacy cycle entirely, building systems that evolve continuously rather than freezing until crisis forces change. Chapter 17 looks beyond government to broader implications and calls for building the community of practice that can make specification-driven development the norm.

This book is written for anyone responsible for or interested in legacy system modernization, but especially for those working in and around government. It assumes familiarity with the challenges of government technology but does not require deep technical expertise. The goal is to provide both a conceptual framework for thinking about legacy modernization in a new way and practical guidance for putting that framework into action.

The legacy crisis in government technology has been building for many decades. Over that time, modernization projects have failed spectacularly and repeatedly. The same patterns have produced the same outcomes. This book argues that the emergence of AI coding assistants creates an opportunity to do something different: to preserve knowledge before it disappears, to verify understanding with the people who actually know how systems work, and why, and to build technology that can evolve rather than accumulating decades of complexity.

The window to address the urgent problem we face is open. The tools to do things differently exist. What remains to be seen is whether we as a community of technologists in and around government will rise to the challenge.

Part I

The Legacy Modernization Crisis

"Diseases desperate grown, by desperate appliance are relieved, or not at all"

Hamlet; Act 4, Scene 3

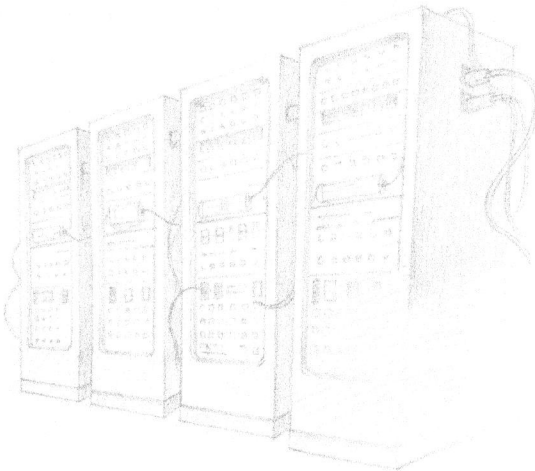

Chapter 1: The Quiet Crisis

In early April 2020, as COVID-19 began its relentless spread across the United States, New Jersey Governor Phil Murphy stepped before television cameras to make an unusual and urgent plea. The state needed volunteer programmers. Specifically, programmers who knew the COBOL programming language.

COBOL, or more formally the **CO**mmon **B**usiness-**O**riented **L**anguage, was first adopted in 1959 when Dwight Eisenhower occupied the White House and the Soviet Union launched *Luna 1* toward the moon.[1] New Jersey's unemployment insurance system, suddenly flooded with claims as businesses shuttered and layoffs mounted, was running on this 60-year-old technology. The system was collapsing, and New Jersey needed people who could keep the aging code functioning long enough to process the wave of desperate claims.

This was not an isolated incident. Across the country, state after state discovered that the systems processing unemployment benefits—systems that millions of Americans suddenly, urgently needed—were built on technology that predated the personal computer. In California, officials were confronting not just aging technology but also billions in fraudulent claims that had exploited relaxed eligibility rules.[2] In the wake of the pandemic, Kansas officials lamented that despite unemployment returning to low levels, the state was "still using antiquated equipment" from the 1970s.[3]

The pandemic exposed what many in government technology have known for years: America's digital infrastructure, the systems that underpin everything from unemployment insurance to tax processing to veterans' healthcare, is alarmingly fragile. Not fragile in the way a new technology might have bugs to work out, but fragile in the way a bridge built in 1960 and never properly maintained becomes structurally unsound. These systems will eventually fall over. The critical question this raises for those of us in the technology community is: "what happens when they eventually do?"

The Scale of Legacy Systems in Government

The federal government spends more than $100 billion annually on information technology and cyber-related investments.[4] That figure represents one of the largest IT budgets in the world, funding the technology that processes tax returns for every American household, manages benefits for tens of millions of people, supports the healthcare of millions of our nation's veterans, and enables countless other functions of government to operate.

Where does that money go? About $83 billion of it (a whopping 79 percent) goes to operations and maintenance of existing systems.[5] Not new development. Not innovation. Not building the capabilities that citizens increasingly expect from their interactions with their government. Nearly four out of every five IT dollars spent simply keeps old systems running.

This is the spending paradox at the heart of government technology. We invest enormous sums, but those investments mostly flow to maintaining what we already have rather than building what we need now, or will need soon. The systems consuming this maintenance budget aren't just old. They're obsolete in ways that create compounding problems.

When government officials and watchdogs talk about "legacy systems," they're describing technology with specific, measurable characteristics. These systems use programming languages that were considered outdated decades ago. They run on hardware and software that vendors no longer support, meaning security patches don't exist and replacement parts must be scavenged or custom-made. They operate with known cybersecurity vulnerabilities that can't be fixed without replacing the entire system. And they depend on a shrinking pool of experts who understand how they work.[6]

Annual Federal IT Spending	Critical Legacy Systems	The COBOL Footprint
Total: Over $100 billion annually	11 federal systems identified as most in need of modernization	250 billion+ lines of COBOL code running daily worldwide
Operations & Maintenance: $83 billion (79%)	Age range: 23 to 60 years old	60 million lines at Social Security Administration alone
New Development & Modernization: $22 billion (21%)	Annual operating costs: $754 million collectively	95% of ATM transactions processed using COBOL
	8 of 11 use outdated programming languages	43% of banking systems rely on COBOL
	7 of 11 have known cybersecurity vulnerabilities that can't be fixed without replacement	

In 2025, the Government Accountability Office identified 11 federal legacy systems most in need of modernization. Eight use outdated programming languages. Four have unsupported hardware or software. Seven operate with known cybersecurity vulnerabilities that can't be remediated without modernization.[7] These aren't experimental systems or minor applications. They range from 23 to 60 years old and collectively cost $754 million annually just to operate and maintain.[8]

Consider what that age range means. The youngest of these critical systems was launched when personal computers were just becoming common in offices. The oldest dates to an era when punch cards were still standard and the internet was science fiction.

Two separate Treasury Department systems identified by GAO run on COBOL and Assembly Language Code—programming languages that have a dwindling number of people available with the skills needed to support them.[9] The Environmental Protection Agency operates a system with obsolete hardware that manufacturers no longer support and known cybersecurity vulnerabilities that can't be fixed without replacement.[10] The Indian Health Service uses a system originally implemented in 1969, based on the MUMPS programming language, which reached its last standard form 25 years ago and is considered both legacy and obsolete.[11]

These are not edge cases. The Social Security Administration maintains systems with 60 million lines of COBOL code in production, supporting programs that provide benefits to tens of millions of Americans.[12] When the IRS processes your tax return, it runs through the Individual Master File, a system with components written in Assembly language dating to the 1960s.[13]

The ubiquity of COBOL in government and financial systems tells its own story. More than 250 billion lines of COBOL code run daily in production systems worldwide.[14] The language processes 95 percent of ATM transactions and supports 43 percent of banking systems.[15] At least 45 states still run systems built on COBOL.[16] This is the technological foundation supporting trillions of dollars in daily commerce and the delivery of essential government services.

The People Who Remember

Tom had worked at the agency for 32 years. He knew the payroll system inside and out—not just how to use it, but how it actually worked. Where the data transformations happened that no one else understood. Which business rules were embedded in which subroutines. Why certain calculations produced seemingly odd results that were actually correct implementations of complex policy requirements.

When Tom retired, he spent two weeks trying to document what he knew. But how do you capture three decades of accumulated understanding in 80 hours? The agency threw him a party, gave him a plaque, and wished him well. Six months later, a problem emerged in a calculation that affected thousands of employees. No one could figure out what the code was supposed to do. Tom came back as a consultant. He'll be 70 next year.

This scenario is fictional, but the pattern it represents plays out constantly across government agencies. The workforce that built the systems governments run on is retiring, and the knowledge required to maintain them is walking out the door. Every day.

The demographics are pretty stark. The average COBOL programmer is between 50 and 70 years old—roughly the same age as the language itself.[17] By 2030, every Baby Boomer will be at least 65 years old. Approximately 10,000 Boomers are expected to retire each day until then.[18] The federal workforce skews significantly older than the private sector: the average federal worker is 47.2 years old, compared to 42.2 for the overall U.S. labor force.

Category	Aspect	Details
COBOL Overview	Full Name	COmmon Business-Oriented Language
	Created	1959
	Original Purpose	Make programming more accessible by using English-like syntax for business applications
Why COBOL is Still Used	Reliability	Extremely reliable for processing large volumes of transactions
	Precision	Handles financial calculations with precision
	Maturity	Systems written in COBOL have been tested and refined over decades
	Risk Aversion	Replacing working systems is expensive and risky
The Problem	Educational Gap	Few universities teach it anymore
	Career Concerns	Younger programmers avoid learning it
	Workforce Crisis	The workforce that built these systems is retiring
	Rising Costs	Costs to maintain COBOL systems increase 10-15% annually as expertise becomes scarce

Over 28 percent of full-time permanent federal workers are age 55 or above, compared to about 23 percent in the private sector. Only about 7 percent of federal employees are under 30, compared to nearly 20 percent in the general labor force.[19] For specialized roles maintaining legacy technology systems, the age disparity is even more pronounced.

This wouldn't be a crisis if knowledge were being transferred to younger staff. But it isn't. COBOL is taught at only a handful of U.S. universities.[20] The average age of a COBOL programmer is 58, and 10% are retiring each year. A 2019 estimate, predicted that there would be 84,000 unfilled Mainframe COBOL programming positions the following year [21] Younger programmers actively avoid learning legacy programming languages, reasonably viewing them as career dead ends in a field where modern skills command higher salaries and more interesting opportunities. None of this bodes well for legacy government technology systems.

Andrew Starrs, a technology officer at Accenture, put it bluntly in testimony before the House Ways and Means Committee: "It's not so much that an individual [COBOL programmer] may have retired, [they] may have expired, so there is no option to get him or her to come back."[22]

When these systems break—and, as evidenced by the experience of New Jersey and other states during the Pandemic, they do break—agencies find themselves calling retirees, offering consulting contracts to people who left years ago. Firms like *COBOL Cowboys* maintain networks of veteran programmers, connecting agencies with retired experts who can parachute in for emergency fixes.[23] During the pandemic unemployment surge, multiple states brought back retired programmers to help keep failing systems operational.[24] This works as a temporary stopgap. It is not a solution.

What COBOL Actually Looks Like

```
IDENTIFICATION DIVISION.
     PROGRAM-ID. MainProgram.

     DATA DIVISION.
     WORKING-STORAGE SECTION.
     01  USER-CHOICE        PIC 9 VALUE 0.
     01  CONTINUE-FLAG      PIC X(3) VALUE 'YES'.

     PROCEDURE DIVISION.
     MAIN-LOGIC.
         PERFORM UNTIL CONTINUE-FLAG = 'NO'
             DISPLAY "------------------------------"
             DISPLAY "Account Management System"
             DISPLAY "1. View Balance"
             DISPLAY "2. Credit Account"
             DISPLAY "3. Debit Account"
             DISPLAY "4. Exit"
             DISPLAY "------------------------------"
             DISPLAY "Enter your choice (1-4): "
             ACCEPT USER-CHOICE

             EVALUATE USER-CHOICE
                 WHEN 1
                     CALL 'Operations' USING 'TOTAL '
                 WHEN 2
                     CALL 'Operations' USING 'CREDIT'
                 WHEN 3
                     CALL 'Operations' USING 'DEBIT '
                 WHEN 4
                     MOVE 'NO' TO CONTINUE-FLAG
                 WHEN OTHER
                     DISPLAY "Invalid choice, please select 1-4."
             END-EVALUATE
         END-PERFORM
         DISPLAY "Exiting the program. Goodbye!"
         STOP RUN.
```

A simple COBOL program. The verbose, English-like syntax of COBOL was a feature in 1959 when it was created. Today it seems archaic when compared to modern programming frameworks and coding conventions.

What walks out the door with retiring experts isn't just technical knowledge about programming syntax. It's institutional understanding of why systems work the way they do. Why is this calculation performed in this specific sequence? "Because of a policy change in 1987 that required a workaround for a data structure limitation." Why does this field have this validation rule? "Because of an audit finding in 1993." Why does this batch process run at 2:00 AM every day? "Because that's when a now-obsolete mainframe had available capacity."

Most of this knowledge is not documented. When you ask why a system does something in a particular way, often the only answer is "that's how it's always been done" or "Tom knew, but he retired 3 years ago." The business rules that govern billions of dollars in payments, the edge cases that trigger specific behaviors, the rationale behind code written before today's senior managers joined the workforce—all of it exists primarily in the heads of people who are leaving, or soon will.

As experts retire, maintenance costs increase. The Office of Personnel Management projected that costs for maintaining its mainframe retirement systems would rise 10 to 15 percent annually as personnel with necessary coding expertise retired since they can't be easily replaced.[25] The remaining staff spend more time on maintenance and less on improvement. Knowledge gaps create risk. Risk creates caution. Caution prevents change. The problem feeds on itself.

A Graveyard of Grand Plans

People that work in government technology, and many people outside this community, have known about this problem for decades. The response has typically been to try to fix it with large-scale modernization projects. The results speak for themselves.

From 2003 to 2012, only 6.4 percent of federal IT projects with $10 million or more in labor costs were successful.[26] More than half of large federal IT projects are delayed, over budget, or fail to meet expectations. Over 40 percent are judged complete failures.[27] IT acquisitions and operations have been on GAO's High-Risk List since 2015, a designation reserved for government operations vulnerable to fraud, waste, abuse, and mismanagement.[28]

The risks of "big bang" deployments in legacy modernization—attempting to replace entire systems at once rather than incrementally—are starkly illustrated by Canada's Phoenix payroll system. In 2009, the Government of Canada set out to replace a forty-year-old payroll system serving 290,000 federal employees across 101 departments. The plan seemed straightforward: implement commercial software, consolidate pay processing from 46 distributed centers into a single location, and save $70 million annually.

When Phoenix launched in February 2016, it caused pay problems for 80 percent of federal employees within months.[29] Some workers went months without pay. The project that was supposed to save money has cost Canadian taxpayers more than $5 billion, with no end in sight.[30] The complexity was immense—encompassing more than 80,000 pay rules across 105 collective bargaining agreements—but when the vendor engaged to implement the new system estimated it would cost $274 million against a budget of $155 million to do the work properly, executives removed or deferred over 100 pay processing functions rather than seek additional funding. They compressed timelines, cancelled pilots, and launched anyway.[31]

What made Phoenix particularly instructive was not just that it failed—large government IT projects fail routinely—but that it failed as a textbook case of legacy modernization gone wrong. This wasn't an effort at building something new; it was replacing something that worked with something that didn't. Canada's Auditor General called it "an incomprehensible failure of project

14

management and oversight."[32] The failure pattern exemplifies what happens with big bang approaches to legacy system replacement: political deadlines overriding technical readiness, inadequate testing of complex business rules, unclear governance, and insufficient contractor oversight. Most critically, the project had extensive requirements documents but no verified specification of what the legacy system actually did and why. The institutional knowledge of experienced pay advisors—people who understood the quirks and workarounds that made the old system function—was eliminated along with their positions. Nearly a decade later, more than 370,000 pay disputes remain outstanding.

The Department of Veterans Affairs has spent nearly two decades and billions of dollars trying to modernize its electronic health record system. Four separate efforts. The first three were abandoned due to concerns about project planning, costs, and length of time to deliver capabilities.[33] The current effort, which began in 2017, first deployed in October 2020. By April 2023, after deploying to five medical centers, VA paused further deployment because veterans and clinicians reported the new system wasn't meeting expectations.[34] Despite some improvements, users at the initial sites remain generally dissatisfied with the system.[35] Life cycle costs are estimated at $49.8 billion.[36] The system still isn't functioning as intended.

These aren't outliers. They're the pattern. States have their own graveyards of failed modernization projects. California's DMV canceled its modernization program in 2013 after seven years and $134 million spent—the second attempt at modernization after a previous $44 million failure in 1994, also cancelled after seven years.[37] Both attempts suffered from project mismanagement, overly optimistic schedules, and risks that were never effectively managed. The lack of a modern system led to long customer lines, incorrect voter registrations, system outages, and became a campaign issue in the state's gubernatorial election.

Oregon has had more success, completing a massive technology overhaul that retired nearly 100 legacy systems and moved them to modern technology.[38] Texas is early in planning its DMV transformation, requesting funding to develop requirements and define the project.[39] Missouri is consolidating decades-old systems, with staff supporting the legacy infrastructure preparing to retire.[40] The stories vary in outcome but share common threads: aging systems requiring specialized knowledge, staff retirements creating urgency, and enormous complexity that makes change risky.

The common threads across failures are remarkably consistent. Projects try to do too much at once in big bang approaches. Requirements shift throughout development. Governance is unclear, with confusion about decision authority. Oversight of contractors is insufficient. Testing gets compressed or skipped as political deadlines loom. Technical readiness becomes secondary to schedule commitments made years earlier.

The pattern persists despite hundreds of GAO recommendations, despite high-profile failures, despite executive orders and policy reforms.[41] GAO has made more than 1,800 recommendations to the Office of Management and Budget and federal agencies since 2010 aimed at improving their management of IT. [42] The fundamental dynamics that produce failure haven't changed.

Between a Rock and a Hard Place

This history creates a dilemma that anyone involved in government technology understands viscerally. The cost of keeping legacy systems running continues to escalate. But the demonstrated risk of trying to replace them is enormous.

As systems age and expertise retires, costs compound. Security vulnerabilities accumulate, creating compliance risks and the potential for breaches that could compromise sensitive data about millions of people. Systems become less able to adapt to new requirements, as the pandemic unemployment surge brutally demonstrated. Citizens experience degraded service—long wait times, limited online options, systems that fail during peak demand.

The IRS alone spent about $39 million in fiscal 2024 maintaining legacy systems that are slated for retirement—systems that agency officials acknowledge are unsustainable but can't yet be replaced.[43] And that's for systems the agency has plans to retire. The cost of maintaining systems that will remain in service for years compounds annually.

But the risk of attempting modernization is equally real. Careers have been destroyed by failed modernization projects. Political capital gets consumed. Mission-critical systems that can't afford downtime during transition periods sit in a precarious state. Complexity that defies comprehensive requirements capture means that even well-intentioned efforts can miss critical functionality or introduce new bugs that affect millions of people.

Individual decision-makers make rational choices in this environment. A failed modernization is career-ending. Continued maintenance is just business as usual. Budget cycles reward showing what you did this year, not investments that pay off in three to five years. Political leadership turns over; modernization timelines don't align with election cycles. The incentive structure produces paralysis.

Meanwhile, the problem gets worse. The government workforce continues to get older. Knowledge continues evaporating. Security vulnerabilities multiply. Maintenance costs grow. The eventual modernization, looming like a specter somewhere in the future, becomes harder and more expensive.

This is the trap: waiting guarantees the problem will get worse, but moving forward has historically meant flirting with disaster. The rational response to this bind has been incremental investment just sufficient enough to keep systems operating, with occasional attempts at modernization that often end in expensive, and very public, failure.

Something Different This Time

In the last few years, something has changed that creates genuine opportunity to break this cycle. This is not just another wave of optimism about a new technology that will supposedly solve all our problems. The technology landscape has shifted in ways that make previously impossible approaches newly viable.

AI-assisted development tools have become capable of tasks that until recently required deep human expertise. Large language models can analyze legacy codebases to extract and compile useful information about antiquated systems. They can be used to separate business logic from technical implementation. They can generate documentation and specifications from code. They can translate between programming languages. And they can do this at scale and speed that makes modernization timelines fundamentally different.

This isn't hypothetical. AI coding assistants like Claude, GitHub Copilot, and Cursor are in widespread production use today. Developers use them daily to understand unfamiliar codebases, generate boilerplate code, and translate between languages. The tools have limitations and produce imperfect output that requires human review. But they're useful enough that companies integrate them into standard workflows.

What makes this relevant to legacy modernization is what AI makes newly practical. Direct code translation—running COBOL code through a translator to produce Java or Python—has been possible for years. The problem is that translated code carries forward all the assumptions, structure, and often the bugs of the original code without enabling human verification. It's fast but potentially very dangerous.

What if instead AI could help us understand what systems do before we rebuild them? What if we could extract specifications—descriptions of business logic, calculations, validation rules, workflows—that domain experts could verify? People who don't understand code but who do understand policy requirements could review specifications to confirm accuracy. The specification then becomes the valuable artifact, not any single code implementation.

This changes the verification problem entirely. You don't need a programmer to verify that a benefits calculation specification matches policy requirements. You need someone who understands benefits policy. Domain experts can do what they're experts in: determining whether business rules are correct. The AI assists in extraction and translation. Humans verify correctness against real-world requirements, policies, and statutes.

The specifications then become the source of truth. These specifications can then be used to generate code in modern languages. When requirements change, specifications get updated first, then code. When the next modernization becomes necessary—and it eventually will—specifications provide a verified foundation rather than requiring archeology to understand what existing systems actually do.

This isn't magic. Specifications can be wrong. AI tools can hallucinate. Domain experts can disagree about requirements. The process requires iteration, verification, and honest assessment of where AI helps and where it potentially creates new problems. But it offers something we haven't had before: a path to capturing institutional knowledge at scale before it walks out the door, and converting that knowledge into artifacts that outlive any particular technology implementation.

The workforce crisis is reaching an acute stage. Retirements in the public sector are accelerating. The pandemic demonstrated how fragile these systems are under stress. At the same time, AI capabilities have reached a threshold of usefulness for exactly the kind of knowledge extraction and translation that legacy modernization requires. The question is whether we can mobilize to use these capabilities while we still have domain experts available who can verify the output.

This legacy crisis in government has been building for decades. We've known about aging systems, retiring workforce, and mounting technical debt for a while now. We've watched modernization projects fail spectacularly and repeatedly. The incentive structures that produce paralysis are well understood. None of this is new.

What's new is the possibility of doing something different. Not replacing old code with new code, but extracting and preserving knowledge in forms that humans can verify and machines can translate. Not big bang replacements, but systematic knowledge capture that enables incremental modernization. Not hoping that COBOL experts will magically appear, but building specifications that people with policy expertise can maintain.

The rest of this book explores how to make that shift. Chapter 2 examines how we got here—the history of government IT procurement, the dynamics that created decades of technical debt, and why the mainframe era persists. Chapter 3 dissects why traditional modernization approaches fail and what those patterns teach us about better paths forward.

Then we'll explore what's now possible with AI-assisted specification-driven development, how to preserve knowledge before it disappears, and how to build systems designed to evolve rather than ossify. This is about breaking a cycle that has consumed billions of dollars and limited government's ability to serve citizens effectively.

The quiet crisis isn't quiet anymore. The pandemic made it very, very loud. It's up to us whether we'll use this moment to build something better, or wait for the next crisis to expose how little has changed.

Chapter 2: How We Got Here

In 1959, a committee of computer scientists and business representatives gathered to design a new programming language. Their goal was practical: create something that could run on any manufacturer's computer and that businesspeople, not just mathematicians and scientists, could read and understand.1 They called it COBOL, the **CO**mmon **B**usiness-**O**riented **L**anguage. It was initially designed to be temporary–a stopgap until something better came along.

Sixty-five years later, COBOL processes an estimated 95% of ATM transactions, 80% of in-person transactions, and runs systems at the heart of banking, insurance, and governments worldwide. The Department of Defense system that helped design it has been replaced many times over. The "temporary" language outlasted nearly everything built to succeed it.

This is the central paradox of legacy systems, one that Marianne Bellotti articulates in her book *Kill It with Fire*: legacy systems become legacy precisely because they have worked well over time.2 A system that failed would have been replaced sometime in the past. A system that worked well enough, year after year, became critical infrastructure. Features were built on top of it. Processes adapted to it. Institutional knowledge accumulated around it. And gradually, over time, a successful system became a sort of trap.

Understanding how government IT reached its current state requires acknowledging and understanding this critical paradox. The systems we now struggle to modernize were, for the most part, triumphs of their time. The mainframes that agencies are desperate to replace were once cutting-edge investments that transformed government's capacity to serve citizens and provide services. The COBOL code that now seems like a liability was once a sensible choice that enabled decades of reliable government operation.

Success became a crisis through the accumulation of decisions made over decades, each reasonable in its moment, that together created a situation no one would have designed intentionally.

The Long Shadow of the Mainframe

The contemporary stereotype holds that government is slow to adopt new technology, always years behind the private sector. The history of computing tells a different story. The federal government was not just an early adopter of computing technology; it was a driving force in the industry's development.

The Social Security Administration began using computers in the 1950s, when commercial computing was still nascent. The IRS followed. By the 1960s, agencies across government were investing heavily in mainframe systems to manage the growing complexity of federal programs. Government contracts funded much of the early computer industry's growth.

This history contains an irony worth noting: the government agencies now criticized for running "ancient" technology were, in their time, sort of on the cutting edge. The mainframes that seem hopelessly outdated today were bold investments in transformative technologies. The problem isn't that government was slow to adopt computing. The problem is that adoption was so successful, and the resulting systems so deeply embedded, that moving beyond them became progressively harder.

These early investments made sense. Mainframes offered unprecedented capabilities for processing large volumes of transactions reliably. COBOL, designed specifically for business applications, allowed agencies to encode complex rules into systems that could execute them consistently, millions of times.

The alternative was armies of clerks processing paper forms, with all the errors and delays that approach would have entailed had it been practical to try.

The very success of these early systems created the conditions for the legacy crisis we find ourselves in today. When a system processes tax returns or Social Security benefits reliably for decades, it becomes deeply embedded in agency operations. Staff are trained on it. Procedures are built around it. Other systems are designed to interface with it. The cost of changing it grows with each year it operates successfully.

Bellotti describes this dynamic in detail: systems that survive do so because they deliver value. The survival of these systems itself is evidence of success. But survival also means accumulation. Each year of operation adds another layer of dependencies, another cohort of staff trained on the system, another set of processes that assume the system works as it does.[3]

The programming languages of legacy systems weren't the only product of their time. The architectural approaches embedded in these systems also reflect the paradigms and constraints of the era in which they were built.

Modern software development has embraced principles that were not yet established when most legacy government systems were designed: modularity and separation of concerns, microservices architectures, iterative and incremental delivery, continuous integration and deployment, loosely coupled components that can be modified or replaced independently. These approaches emerged from decades of hard-won experience about what makes software maintainable and extensible.

Legacy systems, by contrast, tend to be large monolithic applications where components are tightly coupled, where a change in one area can have unexpected effects elsewhere, and where the system must be understood as a whole rather than as a collection of independent parts. This wasn't poor engineering at the time. Computing resources were expensive, and the overhead of modular architectures was a real cost. Tight coupling allowed for optimizations that mattered when every CPU cycle counted.

But these architectural choices create specific challenges for modernization. It can be difficult to extract a single component from a tightly coupled monolith and replace it with a modern equivalent. The Strangler Fig pattern—discussed in detail in later chapters of this book—which works well for systems with clear module boundaries, becomes much harder when those boundaries don't exist. The system resists incremental change because it simply wasn't designed for incremental change.[4]

This compounds the knowledge problem. Understanding a modular system means understanding individual components and their interfaces. Understanding a monolithic system means understanding how everything interacts with everything else. The cognitive load is higher, the documentation challenge is greater, and the risk of unintended consequences from any change is more severe.

The specific choices made in the mainframe era continue to shape government IT today. COBOL remains in production not because anyone thinks it's the best language for new development, but because billions of lines of working COBOL code exist and function. The GAO's July 2025 report found that eight of the eleven federal legacy systems most in need of modernization use outdated programming languages, including COBOL and Assembly Language Code.[5]

The Department of Defense contract management system profiled by GAO (designated "System 3" in the report to protect potentially sensitive details), initially developed in 1964, still runs on "obsolete programming languages, specifically COBOL and assembly language code." The last significant system upgrade was in 2005, which "involved relocating the systems mainframe;

however, the underlying technology and architecture remained largely the same."[6]

It would be a mistake to view this history as simply a cautionary tale. The mainframe systems built in this era accomplished remarkable things. They enabled the administration of programs serving hundreds of millions of people. They operated with reliability that modern distributed systems often struggle to match. They processed transactions at scale that would have been impossible manually.

The problem is not that these systems were bad choices. The problem is that good choices made in one technological era create constraints in the next. This may be the single most important insight for understanding the legacy system crisis. The mainframe approach was right for its time. COBOL was the appropriate tool for its era. The centralized, batch-processing architecture made sense given the technology available. Each of these choices was reasonable, even optimal, when made.

But technology evolves, and what was optimal once becomes limiting over time. The very success of these early systems meant that enormous institutional investments were built around them. Changing the foundation meant changing everything built on top of it. The better the original choice, the more was built on it, and the harder it became to move past it.

The Budget Structure That Prevents Continuous Improvement

Government budgeting evolved around physical infrastructure: bridges, buildings, roads. You fund a capital project to build the asset, then you fund a smaller operations and maintenance budget to keep it functioning. The asset depreciates predictably over time, and eventually you fund another capital project to replace it.

This model works reasonably well for physical infrastructure. A bridge built in 1980 largely does the same thing in 2024 that it did when constructed. The core requirements largely don't change. Maintenance typically means fixing what breaks, not adapting to fundamentally new demands. [7]

Software is fundamentally different. Requirements change constantly as policies evolve, user expectations shift, and the security landscape transforms. A system that isn't actively developed isn't staying the same; it's falling behind. But the typical government budget process doesn't reflect this reality.

The consequences of this mismatch are visible in federal IT spending. For fiscal year 2025, approximately $83 billion (79%) of federal IT spending went to operations and maintenance, with only about $22 billion (21%) available for development, modernization, and enhancement.[8]

This ratio tells an important story. Agencies spend the vast majority of their IT budgets just keeping existing systems running. There's little left for improvement, let alone transformation. And because operations and maintenance budgets are designed for maintenance rather than modernization, the spending that does occur goes toward patches and workarounds rather than fundamental improvement.

The budget process creates a perverse dynamic around modernization. Thoughtful, incremental investment that prevents future problems doesn't usually fit the typical project model. There's no compelling way to request funding for "modernization that will pay off in seven to ten years" or "improvements that could prevent the crisis we might otherwise face in 2030."

But emergency funding for a system in crisis? That's a compelling budget request. Agencies have learned that the way to get modernization funding is to have a crisis, or at least to be able to credibly claim one is imminent. The result is that systems are allowed to deteriorate until they reach crisis point, because that's when funding becomes available.[9]

Agency	Total IT Spending	O&M	DME
All CFO Act Agencies	**$105.1 billion**	**79%**	**21%**
Housing and Urban Development	$538 million	97%	3%
Small Business Administration	$318 million	95%	5%
Homeland Security	$10.7 billion	91%	9%
Department of Defense	$29.1 billion	83%	17%
Social Security Administration	$2.2 billion	70%	30%
Department of Transportation	$4.9 billion	60%	40%
Department of the Treasury	$9.5 billion	61%	39%

O&M = Operations and Maintenance; DME = Development, Modernization, and Enhancement
Source: GAO analysis of IT Dashboard data, GAO-25-107795

Note the extremes: HUD at 97% O&M and Small Business Administration at 95% have almost nothing left for modernization, while Treasury (61%) and Transportation (60%) have relatively more flexibility. The overall 79/21 split means that for every dollar spent on building something new, nearly four dollars go to keeping existing systems running.

Congress has acknowledged this problem. The Modernizing Government Technology Act of 2017 created two mechanisms meant to break the cycle. First, it authorized agencies to establish

IT working capital funds, allowing them to retain savings from IT improvements and reinvest those savings in further modernization. Second, it created the Technology Modernization Fund (TMF), a central fund from which agencies could borrow to finance modernization projects, with the expectation that they would repay the fund from the savings their projects generated.

The idea is a good one: give agencies a way to fund modernization outside the normal appropriations cycle, let successful projects fund future projects, and create a virtuous cycle of continuous improvement. Congress initially appropriated $175 million, then added another $1 billion in 2021.

The results, by some accounts, have been modest. As of April 2025, the Technology Modernization Board had approved 69 projects across 34 agencies, totaling over $1 billion. But GAO found that these projects "have thus far achieved minimal cost savings." Of the $756 million in anticipated savings that funded projects had projected, only about $14.8 million had actually been realized.[10] The fund has supported useful work, but it hasn't fundamentally transformed how government approaches IT modernization.

Part of the problem may be the savings model itself. The TMF assumes that modernization produces quantifiable cost reductions: fewer mainframe hours, reduced contractor expenses, lower licensing fees. Some projects do generate these savings. But much of the value from sustained modernization is harder to measure. How do you quantify the savings from incremental improvements that prevent a high-risk big bang replacement ten years from now? What's the dollar value of preserving institutional knowledge before experts retire, or of maintaining a system flexible enough to adapt when policy changes? These benefits are real, but they don't fit neatly into a repayment schedule. A funding mechanism built around measurable savings will undervalue work whose primary benefit is risk avoided or options preserved. The underlying budget dynamics remain largely unchanged.

Private sector software companies treat their products as ongoing concerns. They budget for continuous development, not just maintenance. Roadmaps extend years into the future. Investment in the product continues throughout its useful life.

Government mostly hasn't made this shift. IT systems are still treated as projects with a beginning, middle, and end, not as products requiring sustained investment. The project ends, the development team disperses, and the system enters "maintenance mode," which in practice often means managed decline.

Until budget structures evolve to support continuous investment in IT systems, the crisis cycle will continue. Agencies will maintain systems until they can't, then request emergency funding for crisis-driven replacements, which will themselves enter the same cycle of decline.

The Procurement System That Favors Large Contracts

Federal procurement regulations exist for good reasons. They're designed to ensure taxpayer money is spent wisely, to prevent favoritism and corruption, and to give all qualified vendors a fair chance to compete. These are legitimate goals, and the rules that implement them are not arbitrary bureaucratic obstacles.

But rules designed for one context can create problems in another. Procurement rules evolved primarily around physical goods and construction projects, where requirements can be specified in advance and where the lowest compliant bid is often genuinely the best choice. Software development doesn't work this way.

The overhead of federal procurement creates a natural bias toward larger contracts. The time and expense required to compete for and manage a federal contract is substantial. This overhead is comparable whether the contract is for $10 million or $100 million, which means the overhead is proportionally much smaller for larger contracts.

The result is that federal IT tends toward large, multi-year contracts with major vendors. Small contracts that might allow for experimentation, learning, and course correction are relatively rare. The system is optimized for predictability and control, not for the kind of iterative development that produces good software.

When government agencies contract for software, they typically require extensive customization to meet their specific needs. This often makes sense from the agency's perspective: they have unique requirements driven by law and policy, and commercial off-the-shelf solutions rarely meet all of them.

But customization creates dependencies. Once an agency has invested millions in customizing a vendor's product, switching to a different vendor means losing that investment and starting over. The vendor knows this, which affects the dynamics of the relationship. Over time, agencies can become effectively locked into vendors even when the relationship isn't serving them well.[11]

Long-term vendor relationships create a knowledge asymmetry that further entrenches dependencies. After years of maintaining and modifying a system, the contractor often understands it better than the agency does. This makes the agency dependent on the contractor not just for continued development, but even for understanding what the system does and how it works.

Government contracts typically require documentation and knowledge transfer, but the reality often falls short. The institutional knowledge resides with the contractor's staff, not the agency's. When contracts change hands or when agencies try to bring work in-house, this knowledge gap becomes a major obstacle.

Why Working Systems Become Untouchable

When a system processes tax refunds for millions of people, determines eligibility for benefits that families depend on, or dispatches ambulances to emergencies, the cost of failure isn't measured in lost revenue or customer complaints. A child welfare caseworker who can't access case history might miss a warning sign. A 911 system that goes down for even a few hours could cost lives.

The GAO's 2025 report notes that the eleven legacy systems most in need of modernization "are essential to government operations such as health care, critical infrastructure, tax processing, and national security."[12] These aren't marginal systems that could be taken offline for a few months while a replacement is tested. They're the systems that government runs on.

The systems most in need of modernization are often the ones most difficult to modernize, precisely because they're so critical. A marginal system that nobody depends on can be replaced with relatively low risk. A system at the heart of agency operations carries enormous risk in any transition.

And modernization itself introduces the very risks you're trying to eliminate. The reason to modernize is often to reduce risk: the old system is fragile, hard to maintain, dependent on retiring expertise. But the act of modernization creates new risks: data migration can corrupt records, new interfaces can introduce errors, staff have to learn new workflows during the transition.

Leaving a legacy system alone may feel safer because nothing changes today. But the risks compound over time. The experts get closer to retirement. The hardware gets harder to source. The security vulnerabilities accumulate. At some point, the system will fail or become impossible to maintain, and the crisis will be worse for having been deferred.

The Organizational Dynamics That Embed Technical Debt

Bellotti makes a crucial observation: you can't understand a legacy system purely through its code. The system reflects the organization that built it. Conway's Law, first articulated in 1967, states that organizations design systems that mirror their communication structures. This applies not just to initial design, but to decades of modifications and workarounds.[13]

Messy code often reflects messy organizational boundaries. Duplicated functionality may trace back to turf battles or poor communication between departments. Mysterious workarounds encode policy decisions that were never properly documented. The technical debt isn't just technical; it's organizational history frozen in code.

Legacy systems accumulate layers of modifications over decades. Each layer made sense at the time. A policy change required a quick fix. A bug was patched rather than properly repaired because the expert who understood that module had left. A new requirement was bolted on because there weren't sufficient resources budgeted for a proper redesign.

These individual decisions, each rational in its moment, accumulate into systems that are difficult to understand and dangerous to modify. The current behavior of the system is the product of thousands of decisions by hundreds of people over decades, many of whom are no longer available to explain their reasoning.

When asked why certain calculations work the way they do, the honest answer is often "because that's how they've always worked." The original requirements documents, if they exist at all, describe what the system was supposed to do in 1987, not what it actually does in 2026. The code is the only authoritative record of the system's behavior, but the code doesn't explain why.

This documentation gap is both cause and consequence of the legacy problem. Without good documentation, modifications are risky, so people avoid them. Because modifications are avoided, there's no occasion to update documentation. The gap widens with each passing year.

Bellotti argues that technical debt is often better understood as organizational debt. The state of the codebase reflects the state of the organization over time. Periods of underfunding show up as accumulated shortcuts. Reorganizations appear as inconsistent interfaces between modules. High turnover manifests as lost institutional knowledge and abandoned documentation.[14]

This reframing matters for modernization. If technical debt is purely technical, you might think you can fix it with technical solutions: refactoring, rewriting, better tools. But if technical debt is organizational, you need to address organizational dynamics as well. The same organizational patterns that created the legacy system will create the same problems in its replacement unless something changes.

The People Problem

Code tells you what a system does, but not why. The reasons behind design decisions, the edge cases that prompted specific logic, the policy interpretations embedded in calculations—these live in the minds of the people who built and maintained the system, not in the code itself.

When those people leave, the knowledge leaves with them. And unlike code, which at least persists even if it's poorly documented, human knowledge simply disappears. A system that was fully understood by its maintainers becomes progressively more mysterious as those maintainers retire.

The federal workforce is older than the American workforce overall, and the gap is widening. The average federal worker is 47.2 years old, compared to 42.2 for the overall U.S. labor force. Over 28 percent of full-time permanent federal workers are age 55 or above, compared to about 23 percent in the private sector. Only about 7 percent of federal employees are under 30, compared to nearly 20 percent in the general labor force.[15] For specialized roles maintaining legacy systems, the age distribution is even more pronounced—the average COBOL programmer is between 50 and 70 years old. By 2030, all Baby Boomers will be at least 65.

At agencies with critical legacy systems, this isn't an abstract demographic trend. It's a countdown. The people who understand how the systems work are leaving, and there aren't enough people being trained to replace them.

For DOD's contract management system, the GAO found that "the average age of developers and technical subject matter experts on the system's team is above 60, putting the system at significant risk in being able to support and maintain it into the future." Officials stated that "it is difficult to find COBOL and assembly language code developers and the learning curve once they are identified is also significant."[16]

Agencies know this is a problem. Exit interviews are conducted. Documentation is requested. Knowledge transfer plans are written. But the reality rarely matches the plan. Documenting thirty years of accumulated knowledge in someone's final month is impossible. The person retiring may not even know what they know until a question arises that only they can answer.

And the people who might receive that knowledge often aren't available. Younger staff are busy with current demands. There's no slack in the system for the long conversations that real knowledge transfer requires. The documentation that gets produced is often too abstract to be useful when specific problems arise.

The shortage of legacy system expertise has created a market. As noted in Chapter 1, firms like COBOL Cowboys maintain networks of retired programmers who can be called back for consulting engagements. States brought retired programmers out of retirement during the pandemic unemployment crisis. This phenomenon deserves additional attention here because it illustrates a broader pattern: the market response to expertise scarcity.

When a critical skill becomes rare, prices rise, which attracts suppliers. Retired COBOL programmers who might have stayed retired find consulting rates attractive enough to return. Firms emerge to match this supply with demand. The market works, in the sense that organizations that need COBOL expertise can generally find it if they're willing to pay.

But this market solution has limits. The pool of retirees shrinks with each passing year—not just through retirement, but through mortality. The rates they command increase as their scarcity grows. And relying on consultants means the knowledge still doesn't transfer to permanent staff. The organization remains dependent on external expertise, and that expertise is a depleting resource.

Why "Just Rewrite It" Doesn't Work

When a legacy system becomes painful enough, the temptation to start from scratch is powerful. Imagine having a clean codebase, modern technology, good documentation! No more wrestling with forty-year-old COBOL. No more mysterious batch jobs that nobody understands.

This vision is seductive precisely because legacy systems are so frustrating to work with. The appeal of escape is real. But escape attempts fail far more often than they succeed.

The California DMV attempted complete system replacements twice—in 1994 and 2006. Both failed after approximately seven years and tens of millions of dollars. The same pattern repeats across government.[17]

Complete rewrites underestimate the complexity embedded in working systems. Bellotti points out that a system running for decades has been solving problems for decades. All those patches and workarounds represent solutions to real problems that arose in production. Edge cases that nobody anticipated, policy interpretations that required special handling, integrations with other systems that evolved over time.[18]

A rewrite team looks at the existing system and sees a mess. What they're actually seeing is accumulated solutions to accumulated problems. Unless they understand why each decision was made, they'll rediscover the same problems the hard way—in production, under pressure, with users depending on the system.

Rewrites take time. Large government system rewrites often take years. During those years, the existing system continues to evolve. Policies change. New requirements emerge. Bug fixes are applied. By the time the rewrite is ready to deploy, it's targeting a specification that's years out of date.

This creates a dilemma. Do you freeze the legacy system during the rewrite, denying users new capabilities for years? Do you try to keep the rewrite synchronized with ongoing changes to the legacy system, essentially doubling your development burden? Either choice has serious costs.

Frederick Brooks described the "second system effect" in *The Mythical Man-Month*: designers tend to overload their second system with all the features they wished they could have included in the first.[19] Government rewrites are particularly susceptible to this. After years of living with a constrained legacy system, stakeholders see the rewrite as their chance to finally get everything they want.

The result is scope creep that delays delivery, increases cost, and often produces a system that's more complex than the one it replaces. The rewrite becomes its own kind of legacy before it even launches.

Successful legacy modernization typically follows a different pattern: incremental replacement rather than complete rewrite. This approach has its own challenges, but it avoids the worst failure modes of complete rewrites. It allows learning and course correction. It delivers value incrementally rather than all at once. It maintains the accumulated knowledge in the legacy system while gradually reducing dependence on it.

The Accumulation of Reasonable Choices

The legacy system crisis wasn't created by incompetent people making bad decisions. It was created by competent people making reasonable decisions within systems that rewarded certain behaviors and discouraged others. Budget structures that favor big projects over continuous improvement. Procurement rules that favor large vendors over innovative approaches. Organizational incentives that punish failure more than they reward success. Knowledge management practices that never kept pace with turnover.

Each decision made sense in its moment. Investing in mainframes in the 1960s was forward-thinking. Writing COBOL was practical. Building a monolithic architecture was efficient given the constraints. Signing long-term vendor contracts provided stability. Avoiding risky changes to working systems was prudent. Deferring modernization when budgets were tight was understandable.

But the unseen cost of these reasonable decisions accumulated. Each layer of reasonable decisions creates constraints for the next. The mainframes that were cutting-edge became foundations that couldn't easily be changed. The COBOL that was practical became a language few people learn. The long-term contracts that provided stability became dependencies that limited options. The changes that were prudently avoided became changes that were increasingly difficult to make.

Understanding this history matters because it shapes what solutions are possible. You can't solve an organizational problem with purely technical solutions. You can't escape budget dynamics by wishing they were different. You can't modernize systems without reckoning with the knowledge embedded in them.

We can't simply avoid repeating history—the forces that created the current situation haven't fully changed. But we can explore whether new tools and approaches can work within these constraints more effectively than previous attempts. That's what the rest of this book will do.

But first, the next chapter will examine more closely why the traditional modernization playbook—big bang replacements, waterfall development, comprehensive requirements documents—fails so consistently. Understanding the failure modes is essential to understanding why a different approach might succeed.

Chapter 3: The Traditional Modernization Playbook

In 2009, the Government of Canada embarked on what should have been a straightforward modernization project. The federal payroll system was forty years old. It processed pay for 290,000 employees across 101 departments and agencies. The technology was obsolete. Everyone agreed it needed to be replaced.

The plan seemed sensible on paper. Replace the aging legacy system with a commercial off-the-shelf solution. Consolidate pay processing from dozens of distributed centers across the country into a single facility in Miramichi, New Brunswick. Save taxpayers $70 million annually through efficiency gains. The Treasury Board approved $310 million for the initiative. IBM won the contract to implement PeopleSoft payroll software.

Seven years later, in February 2016, the new Phoenix Pay System went live. Within months, the system had caused pay problems for 80 percent of federal employees.[1] Some workers went months without any pay. Others were overpaid and then pursued for repayment years later. Employees who took parental leave or changed positions found their pay calculations hopelessly corrupted. The stress drove some workers to psychiatric care; others lost their homes when they couldn't make mortgage payments.

By the time the Office of Canada's Auditor General completed its investigation, they had coined a phrase to encapsulate what had happened: "an incomprehensible failure of project management and oversight."[2] The project that was supposed to save $70 million annually has now cost Canadian taxpayers more than $5 billion, with no end in sight. Nearly a decade after launch, there are still more than 370,000 outstanding pay disputes.[3]

Phoenix was a case study in dysfunction that extended well beyond the technical approach used for the modernization effort. Political pressure from two successive governments pushed the project forward despite warning signs. Budget constraints led executives to cut essential functionality rather than seek additional funding. A culture of "don't bring bad news to leadership" allowed problems to fester unaddressed. Consultants warned of critical risks; their reports were ignored. These failures of governance, politics, and organizational culture compounded the technical failures.

But underneath all of that dysfunction lies something more fundamental: an approach to legacy modernization that skewed the project toward failure regardless of how well it was managed. Phoenix followed what we can think of as the "traditional playbook" to legacy modernization. Understanding why that playbook can produce failures—even when politics and budgets are not an issue—is essential to finding a better path.

The Big Bang Replacement

The dominant approach to legacy modernization in government involves replacing entire systems at once. The appeal is obvious. Legacy systems are frustrating to maintain and expensive to operate. The temptation is to "just replace it" with something modern, promising a clean break from accumulated technical debt. Stakeholders want a single deadline, a single budget, and a single delivery. The alternative—incremental migration—seems slower and less decisive.

Phoenix exemplified the big bang mentality. The Government of Canada didn't just replace its payroll software; it simultaneously consolidated pay processing from 46 distributed

41

centers into a single location, eliminated 1,200 experienced pay advisor positions, and launched the new system across all 101 departments at once. Any one of these changes in isolation would have been significant. Combining them guaranteed that problems would compound.

Michael Feathers, in his seminal work *Working Effectively with Legacy Code*, offers a definition of legacy code that helps illuminate why big bang replacements fail: he describes legacy code as code without tests.[4] This definition seems narrow at first, but Feathers' insight is critical. Tests serve a crucial purpose: they provide assurance that a system works the way it's supposed to, even when changes are made. Without that assurance, every change is a gamble. You can't safely modify what you can't verify.

The same principle applies at the system level. Without mechanisms to verify that a new system behaves identically to the old one—or more precisely, that it correctly implements the business logic the old system was supposed to implement—you can't safely replace it. The Phoenix team had no such mechanism. They had requirements documents describing desired outcomes, but no verified specification of what the legacy system actually did and why.

The Canadian federal government's payroll rules encompassed more than 105 collective bargaining agreements, each with its own provisions for overtime, shift differentials, leave accruals, and retroactive adjustments. There were more than 80,000 pay rules that needed to be programmed into Phoenix. The system required interfaces with 34 different human resource systems across the Canadian government. To handle this complexity, Public Services and Procurement Canada added more than 200 custom-built programs to the PeopleSoft software.[5]

None of this was adequately tested before launch. When IBM estimated that building Phoenix properly would cost $274 million against an approved budget of $155 million, project executives didn't seek additional funding. Instead, they removed or deferred more than 100 pay processing functions to fit within the budget. They compressed the timeline. They cancelled a pilot that might have revealed the system's readiness issues. They launched anyway.[6]

Martin Fowler and others have long advocated for incremental replacement of legacy systems through patterns like the Strangler Fig pattern, which wraps legacy systems in new components and gradually replaces functionality.[7] Each increment can be tested and validated independently. Risk is distributed across many small changes rather than concentrated in one big launch. But incremental replacement requires something that traditional approaches fail to provide: precise knowledge of what each system component does. You can't safely replace a piece of functionality unless you can verify the replacement works correctly. You can't verify correctness without a specification to verify against.

As discussed in Chapter 2, organizations often choose a big bang approach because budget processes favor large single-year appropriations over smaller multi-year programs. In many cases, career incentives reward launching projects rather than maintaining them, and political timelines create pressure for visible deliverables. "Transformation" sounds a lot better than "incremental improvement." The pain of maintenance feels immediate; the risk of failure feels abstract. And most importantly, without specifications, incremental replacement feels impossible.

The Waterfall Trap

Even organizations that claim to follow agile methodologies often revert to waterfall practices when dealing with legacy modernization. The fundamental assumptions of waterfall can feel safer when the stakes are high. Recall from the previous chapter that many of the systems in government most in need of modernization serve critical functions. Child welfare systems, tax processing systems, emergency dispatch systems-all systems that carry enormous risks when modernization efforts go sideways.

At first glance, waterfall's emphasis on detailed specifications before coding might sound similar to the approach this book advocates. Both involve written documentation that precedes implementation. But the similarity is superficial, and the distinction is crucial.

Waterfall requirements gathering seeks to document a desired future state. These requirements documents capture aspirations, not reality. They end up reflecting what people think they want, filtered through their incomplete understanding of what they currently have.

The approach to knowledge capture detailed in this book is focused on understanding what an existing system actually does, and why. Understanding comes from analyzing working code, verifying behavior with domain experts, and documenting the business logic, quirks, and workarounds that have accumulated over decades. Specifications describe current reality as implemented. They capture what is verified with experts, to enable a modern implementation.

This distinction matters enormously. Waterfall requirements often miss critical functionality because stakeholders don't know how to ask for it—they assume the new system will do what the old one does. When the new system launches without that functionality, everyone is surprised. Phoenix had extensive requirements documents. What it lacked was a verified specification of what the legacy system actually did, and why.

Waterfall appeals to governments because accountability requires documentation, procurement rules require detailed statements of work, and political oversight demands predictability. "We followed the process" offers a shield against criticism even when the process produces disaster.

Phoenix followed a textbook waterfall approach despite the era's increasingly vocal advocacy for agile methods. Requirements were gathered. Contracts were let. The project proceeded through phases with gates and approvals at each stage. Independent consultants were hired to assess readiness. Reports were produced. None of this prevented the eventual disaster.

The Auditor General's investigation revealed why. The governance structure gave Phoenix executives sole authority to decide whether to proceed. Deputy ministers from affected departments had no role in the oversight structure. Independent audit functions that might have raised alarms were not engaged. The process existed on paper, but the people with authority to stop a failing project had insulated themselves from information that would require them to do so.[8]

More fundamentally, the process was oriented toward building a new system, not understanding the old one. The requirements described what Phoenix should do. No one had systematically documented what the forty-year-old legacy system actually did—the thousands of edge cases, all of the informal workarounds, the accumulated policy decisions embedded in code and documented nowhere. That knowledge existed largely in the heads of pay advisors who were about to lose their jobs.

Many organizations adopt agile terminology without agile practices. Sprints become mini-waterfalls. "Working software" becomes "working demonstrations." The big bang is decomposed into phases that still accumulate in a single launch. The fundamental assumption—that we can plan the whole thing upfront—remains unchanged.

The Documentation Fallacy

Organizations consistently promise to document systems before, during, or after modernization. "We'll document the requirements thoroughly before building." "We'll maintain documentation as we develop." "We'll document the system after it's deployed." None of these end up happening in practice. The gap between documentation and reality grows over time.

Feathers observes that legacy systems become "change-resistant" over time as dependencies accumulate and original developers leave. The knowledge of how systems actually work exists in the code itself and in the heads of people who maintain it. Documentation captures intent at a moment in time; the system continues to evolve while the documentation typically remains frozen.

This pattern reflects a documentation parallel to what Feathers calls the "legacy code dilemma"—the observation that you can't safely change code without tests, but you can't add tests without changing code. The documentation version is equally challenging: you can't modernize a system without understanding it, but you can't document what you don't understand, and the people who do understand it are often too busy maintaining it to write anything down.

Phoenix exemplified this pattern. The forty-year-old system being replaced had evolved over decades. Pay rules had been modified to implement one collective bargaining agreement after another. Workarounds had been added to handle edge cases. Institutional knowledge resided largely in the heads of the pay advisors who had spent careers learning the system's quirks. When those advisors were eliminated as part of consolidation, their knowledge walked out the door. No documentation captured what they knew.

Direct Code Translation

AI and automated tools can translate code from one language to another. Legacy code in COBOL or Assembly can be translated to Java or Python. AI tools can now perform these translations at scale. The promise is appealing: maintain functionality while modernizing the technology. Fast, relatively cheap, appears to solve the language obsolescence problem. Bingo!

But direct translation is what Feathers might refer to as "change without understanding." You can translate COBOL to Java without knowing what the COBOL does. The translated code will (hopefully) do the same thing. But "the same thing" includes all the quirks, all the workarounds, all the assumptions that made sense in the past but may not make sense now.

More fundamentally, direct translation doesn't create understanding. You end up with modern code that few people (or maybe no one) can explain, that embeds business logic no one has verified, that will become its own legacy system the moment it's deployed. This approach changes the syntax without changing the situation.

Part III will talk more about the important distinction between transpilation (using AI tools to simply convert legacy code into modern code) and compilation (the process of compiling knowledge and understanding of how a system works).

Manual Rewriting

The alternative to automated translation is manual rewriting by developers who study the legacy system and create new implementations. Experienced developers study legacy code, interview subject matter experts, write new implementations from scratch, and test against legacy system behavior. The result is modern code with documented business logic. This approach produces better understanding than direct translation.

But scaling this approach is a major challenge. A large government system might contain millions of lines of code implementing thousands of business rules accumulated over decades. The number of developers who can understand both legacy code and modern languages is small and shrinking. Even if you could hire enough of them, the timeline would stretch to years. And government doesn't have years. The workforce that understands these systems is retiring now.

Feathers described the sheer difficulty of understanding code written by others. A chapter of his book is titled "I Don't Understand the Code Well Enough to Change It." He offers techniques for building understanding incrementally, but these techniques require time, discipline, and access to the original developers. For legacy government systems, time is short, institutional memory is fragmenting, and the original developers may be retired, or even deceased.[9]

The cost comparison is daunting. Manual rewriting might cost $50-100 per line of code, sometimes more for complex systems. A million-line system at $50 per line means $50 million just for coding. Add requirements gathering, testing, deployment, and training. Total costs quickly exceed available budgets.

Phoenix was essentially a manual rewriting effort. IBM customized commercial software to implement Canadian pay rules. More than 200 custom programs were added. Despite years of work and billions of dollars, the rewriting was incomplete. Functions were deferred or removed to meet deadlines. The result was a system that couldn't do basic things like process retroactive pay.[10]

But something has changed since Phoenix launched in 2016. AI coding assistants have emerged that can analyze legacy code and explain what it does. They can process millions of lines of

COBOL in hours rather than months. They can help generate documentation, identify business rules, and produce initial drafts of specifications.

This doesn't eliminate the need for human expertise—far from it. AI explanations must be verified by domain experts who understand the business context. AI-generated specifications require careful review. But AI dramatically accelerates the understanding phase that has historically made manual rewriting prohibitively expensive.

The equation changes. Instead of needing an army of developers who can read COBOL, you need domain experts who can verify AI-generated explanations. Instead of years spent reverse-engineering business logic, AI can produce a first draft in days. The work of humans becomes verification and refinement rather than initial discovery.

Part II explores this shift in more detail. The economics of knowledge capture are changing. Manual rewriting may have been impossibly slow and expensive in 2016. AI-assisted approaches may make it practical today. And into tomorrow.

The Knowledge Preservation Problem

All traditional modernization approaches share a common failure: they don't solve the fundamental problem of preserving institutional knowledge. Whether you translate code, rewrite it, or replace the system entirely, you lose the knowledge of why the system works the way it does.

Legacy code isn't just syntax. It embeds decades of accumulated decisions. Why does this calculation round down instead of up? Because a collective agreement signed in 1987 specified rounding rules. Why does this workflow have an extra approval step for transactions over $5,000? Because an audit finding in 2003 required additional controls. Why does this system handle leave accruals differently for employees hired before 1995? Because a policy change wasn't applied retroactively.

The Phoenix consolidation effort eliminated pay advisor positions across government and concentrated pay processing in a single location. The pay advisors who lost their jobs had probably accumulated decades of institutional knowledge about the legacy system. They knew the quirks of their departments' collective bargaining agreements. They knew which employees had unusual situations requiring special handling. They knew the workarounds that kept the legacy system functioning despite its age and limitations.

The newly consolidated pay advisors in Miramichi were trained on the new Phoenix system, not on the actual complexity of the federal payroll process. When Phoenix failed to handle situations correctly, they didn't have the institutional knowledge to recognize errors or implement workarounds. The knowledge that made the old system work wasn't transferred to the people running the new system.[11]

Code is a poor knowledge repository. Chapter 5 will discuss why some experts actually view code as a "lossy projection" from a verified software specification. Code captures implementation, not intent. Comments decay and become opaque. Understanding code often requires understanding the context in which it was written. That context is rarely documented.

Feathers writes that to work effectively with legacy code, you have to first understand it well enough to change it safely. His primary mechanism for this is tests—specifically what he calls "characterization tests" that capture existing behavior. When you have comprehensive tests, you can modify code with confidence because the tests will reveal if you've broken something.

Tests & Specifications: Same Principles

Why Tests Are Critical	"Legacy code is code without tests." Tests provide assurance that code changes don't break behavior. Without tests, every modification is a gamble.
Why This Matters for Modernization	Legacy systems are systems without specifications. Specifications provide assurance that replacements implement correct logic. Without specifications, every modernization is a gamble.
Verification Method	Tests are verified by automated execution. Specifications are verified by domain expert review.
Common Purpose	Both preserve knowledge in forms that survive change.

Tests serve as a form of executable specification. They document what the system does in a way that can be automatically verified. They preserve knowledge in a form that survives changes to the underlying code. When a test fails, it tells you exactly what behavior changed. But tests have limitations for modernization. They verify that outputs match expectations, but they don't explain why those outputs are correct.

This is where specifications become important. A verified software specification captures not just what a system does, but why it does it. A specification describes the business rules in language that domain experts—not just programmers—can read and verify. A specification connects code behavior to policy intent.

Think of specifications as serving the same role for modernization that tests serve for maintenance. Tests let you safely modify code because they verify behavior is preserved. Specifications let you safely replace systems because they verify that the replacement implements the correct business logic. Tests are verified by automated execution. Specifications are verified by domain experts who understand the business context. Part III of this book will explore how to create, verify, and use specifications as the foundation for legacy system modernization.

Feathers argues you can't safely change code without tests because you have no way to know if your changes broke something. The same logic applies to modernization: you can't safely replace a system without specifications because you have no way to know if the replacement implements the correct behavior.

A complete specification must capture the business rules the system implements, the policy rationale behind those rules, the edge cases and exceptions, the reasons for apparent inconsistencies, and the historical context that explains current behavior. All of this in language domain experts can verify.

What The Phoenix Fiasco Can Teach Us

The Phoenix pay system remains in operation today, nearly a decade after its troubled launch. The Canadian government has spent years and billions of dollars trying to fix problems that should never have been created. A replacement system is in development, but workers who have tested it report ongoing glitches.[12] The cycle continues.

Phoenix was a failure on many dimensions. Politics played a role: two successive governments pushed forward despite warning signs, neither willing to be blamed for delay. Budget drama contributed: executives cut essential functionality rather than seek additional funding. Organizational culture mattered: a climate where "don't bring bad news to leadership" allowed problems to fester until they were catastrophic. These factors deserve attention, and any large modernization effort must account for political, budgetary, and organizational realities.

But underneath all of this dysfunction lies a more fundamental problem. Even a well-managed version of the Phoenix project—one with adequate funding, reasonable timelines, and leadership willing to hear bad news—would have struggled. Because the traditional modernization playbook doesn't work.

The traditional playbook treats modernization as building a new system to meet requirements. But requirements documents don't capture what legacy systems actually do. The playbook treats testing as something that happens at the end, but you can't test what you haven't specified. The playbook treats knowledge as something that can be transferred through training, but training on a new system doesn't transfer understanding of the old one. The playbook treats a launch as a single event, but big bang deployments concentrate risk.

Phoenix wasn't primarily a failure of execution. It was a failure of approach—the kind of failure that execution excellence can't fully prevent.

But recent changes in the world of technology offer the hope of a new path. AI coding assistants have emerged that can do things that were essentially impossible in 2016. They can analyze legacy code and explain what it does. They can extract business rules from millions of lines of COBOL. They can generate documentation, specifications, and even new implementations. Used correctly, they offer a path out of the modernization trap.

AI can clearly help with legacy modernization. The challenge lies in applying AI capabilities in ways that avoid the failures of the traditional playbook. That requires rethinking what we're actually trying to accomplish. Not simply translating code from one language to another. Not just replacing old systems with new ones. Something more essential: capturing the knowledge embedded in legacy systems in specifications that domain experts can verify, and then using those specifications as the foundation for incremental, validated system modernization.

We'll take up the Government of Canada's Phoenix Pay System example again in Chapter 11: The Politics of Modernization. But first, Part II introduces the AI-assisted development revolution and shows how it makes this new approach possible.

Part II

The AI-Assisted Development Revolution

"O heaven! that one might read the book of fate, And see the revolution of the times"

Henry VI; Part 2 - Act 3, Scene 1

Chapter 4: The Emergence of AI Coding Assistants

Sometime in mid-2017, I woke up with pain at the base of the middle finger on my left hand. For a while I attributed the pain to exercise or general overuse. After all, I told myself, I had spent many years in my youth as a competitive gymnast, regularly abusing my hands on events like the rings and parallel bars. Surely that had something to do with it. But as the pain spread to other fingers and eventually to my other hand, I came to realize that the root cause wasn't in my past. It was sitting right in front of me: my keyboard.

I've always been a hard typist, and I haven't always used proper form. After many years of undisciplined typing, I found myself in a place where every keystroke hurt. As a civic technologist, my hands were one of my most important tools. I had written code and text for my entire professional career, and now the fundamental act of my work caused me pain.

I thought a lot about a quote I had read from The Edge, the guitarist for U2, who once said that "notes are expensive—you don't just throw them around." I came to appreciate how that applies to typing. Keystrokes are expensive, but you don't pay the bill right away. The cost accrues over many years. And eventually the bill comes due. Mine certainly did.[1]

Despite this, when GitHub first announced Copilot in June 2021, I was initially a little skeptical. I had heard promises over the years about tools that promised to make coding easier, or even obsolete. I'd seen visual coding environments in the early 2010s and low-code platforms in the late 2010s. Each promised to revolutionize software development. Each eventually disappointed, at least in my opinion. My hands still hurt. Another tool claiming to transform how we write code? I had heard this story before.

But something unexpected happened as these AI coding tools matured. I started to realize that the skills I had developed over decades working in and around government turned out to be exactly what these tools required. Not typing speed. Not memorization of syntax. Not even deep expertise in any particular programming language. What matters most is the ability to explain things clearly. To break down complex ideas into smaller, discrete chunks. To describe what something should do in precise, unambiguous terms.

These were skills I had honed through years of writing about government technology, explaining software and IT systems to non-technical stakeholders, and helping agencies understand their own processes. Voice transcription tools like Superwhisper had already changed my relationship with text, allowing me to (mostly) write words without the physical cost of typing.[2] I found that using these voice transcription tools further sharpened the way that I articulated thoughts or ideas. I had to be clear, succinct, direct for these tools to work well. As it happens, these are critical skills for using AI coding tools well.

The act of doing this work had resulted in a painful condition in my hands, but–ironically–it had also prepared me for a whole new way of doing the work. My mind was blown.

This personal discovery reflects a broader shift in what modern software development requires. AI coding assistants don't just change how we write code. They change what skills matter most. And for those of us who have spent careers explaining complex systems to other humans, that shift is surprisingly favorable. And even fun.

This chapter explores what these tools are, what they can and can't do, and why they matter particularly for the legacy modernization challenge that Part I of this book described. The goal here is an honest assessment, neither dismissive nor breathlessly enthusiastic. I expect that you'll be as skeptical initially, just as I was. These tools represent something genuinely new, but understanding their limitations is just as important as appreciating their capabilities.

A Rapid Evolution

To understand where we are, it helps to understand how quickly we got here. GitHub announced Copilot as a technical preview in June 2021. This was before ChatGPT existed, before most people had heard the term "generative AI," and before the current wave of excitement about artificial intelligence had begun. Copilot was powered by OpenAI's Codex model, itself derived from GPT-3, and its initial capabilities amounted to what many people called "autocomplete on steroids": suggesting code snippets, sometimes entire functions, based on context and comments.[3]

The reception was mixed. There was excitement about the potential, skepticism about the quality, and concerns about copyright given that the model had been trained on publicly available code in GitHub repositories. Early accuracy was modest. GitHub reported that Copilot correctly autocompleted Python function bodies 43% of the time on the first try, and 57% of the time after ten attempts.[4] Useful, perhaps, but hardly the end of programming as we knew it.

Then the field accelerated. Quickly. Copilot became generally available in June 2022. ChatGPT launched in November 2022, demonstrating conversational AI to a much broader audience and sparking intense interest in what these generative AI models could do. Anthropic released Claude in March 2023.[5] Cursor, Aider, and other AI-native development environments emerged.

The competitive pressure drove rapid improvement across all tools.

By 2024 and 2025, the capabilities had expanded dramatically. Context windows grew from thousands of tokens to hundreds of thousands, allowing models to process much larger codebases at once. Model capabilities improved across coding benchmarks. The architecture evolved from single models to multi-model systems, with GitHub Copilot eventually supporting models from multiple providers including Claude, Gemini, and GPT-4.

Perhaps most significantly, these tools developed what practitioners call "agentic" capabilities. Rather than simply suggesting the next line of code, they could execute multi-step tasks: analyzing a codebase, identifying issues, implementing fixes, running tests, and iterating based on results. Claude Code launched in February 2025 as a command-line tool for this kind of agentic coding work.[6] By November 2025, Claude Opus 4.5 was achieving state-of-the-art results on enterprise coding tasks, with the ability to maintain context and consistency through thirty-minute autonomous coding sessions.[7]

A Brief Timeline of AI Coding Tools

Date	Release
June 2021	GitHub Copilot technical preview
November 2022	ChatGPT launches
March 2023	Claude 1 released
March 2024	Claude 3 family released
June 2024	Claude 3.5 Sonnet with Artifacts

Date	Release
October 2024	Multi-model support in Copilot
February 2025	Claude 3.7 Sonnet and Claude Code
November 2025	Claude Opus 4.5

This trajectory matters because it establishes that these tools have moved from interesting experiments to production-grade capabilities in a remarkably short period of time. What seemed speculative in 2021 is now in widespread use. GitHub reports that developers have used Copilot to accept more than three billion code suggestions, and that every month Copilot contributes to 1.2 million pull requests directly inside GitHub.[8]

We are still early on this curve. The pace of improvement has been faster than most people (certainly me) predicted, and there's no obvious reason to expect it to slow down. But it's equally important not to extrapolate too aggressively. What these tools can do today is impressive. What they might do tomorrow is not yet known.

What These Tools Do

Understanding what AI coding assistants actually do is essential for using them effectively and recognizing their limitations.

At their core, these tools are large language models trained on vast amounts of text, including enormous quantities of source code. They perform pattern recognition and completion at a sophisticated level. When you provide context, whether that's existing code, comments inside that code, or natural language descriptions, the model predicts what should come next based on patterns it learned during training.

This is most definitely not "understanding" in the human sense. The model doesn't know what code does in the way a human programmer knows. It doesn't have a mental model of execution, memory, or state. What it has is an extraordinarily refined sense of what code **looks like**: what patterns typically follow other patterns, what structures tend to appear in what contexts, what naming conventions and idioms characterize different languages and frameworks.

This distinction matters because it explains both the capabilities and the limitations of these tools. The model can produce code that looks correct because it has seen millions of examples of correct code. It can also produce code that looks correct but isn't, because looking correct and being correct are two different things.

Within that frame, these tools offer several core capabilities. Code completion and generation is the most visible capability. The tool suggests the next line, function, or block of code based on context. You can generate code from natural language descriptions, create boilerplate and test cases, and produce documentation. Quality varies significantly depending on context, language, and how well the task matches patterns in the training data.

Code explanation and analysis may be even more valuable for legacy modernization. These tools can translate code into natural language descriptions, identify patterns and dependencies, summarize what complex functions do, and explain unfamiliar code in understandable terms. When you paste a function written in an unfamiliar language into an AI assistant and ask what it does, you often get a remarkably accurate explanation.

Refactoring and transformation capabilities allow these tools to suggest improvements to existing code, convert between styles or patterns, translate between programming languages, and modernize syntax while preserving functionality. This is where the legacy modernization potential of these tools becomes really clear.

Contextual assistance rounds out the picture. These tools can answer questions about codebases, help navigate unfamiliar frameworks, suggest solutions to specific problems, and act as knowledgeable collaborators throughout the development process.

The shift in how practitioners describe these tools is pretty telling. When Copilot first launched, the framing was as an "AI pair programmer," suggesting a tool that works alongside you. By 2025, GitHub was using the phrase "peer programming," suggesting something closer to a colleague. The tools haven't replaced programmers, but they have changed the nature of the collaboration.[9]

An Honest Assessment of Limitations

Overselling AI capabilities undermines trust and leads to poor decisions. Any honest account of these tools must reflect what they can't do well and where they can fail.

Hallucinations are real and they can be dangerous. AI coding assistants will sometimes confidently generate code that doesn't work. They will sometimes invent API endpoints that don't exist. They can misunderstand business logic and produce plausible-looking but incorrect output. The confidence of the output doesn't always correlate with accuracy. A function that looks perfectly reasonable may contain subtle bugs that only become apparent in edge cases that the model (and the person that prompted it) never considered.

One team working on a COBOL migration described their early experiments with AI as producing "a good mix of educated guesses (from us and the model) and hallucinational gibberish."[10] That's an honest assessment. The tools can be remarkably helpful, and they can be remarkably wrong, sometimes in the same conversation.

Common failure modes include misunderstanding complex logic, missing edge cases, making incorrect assumptions about data, and confusing technical implementation with business intent. The model might generate syntactically correct code that is semantically wrong. It might produce verbose or inefficient solutions when more elegant ones exist. It might confidently assert things about your codebase that simply are not true, or only partially so.

Context limitations remain significant even as context windows have expanded. Early tools could only consider a few thousand tokens at a time. Modern tools can handle hundreds of thousands. But complex legacy systems often exceed even these expanded limits. Information at the edges of the context window may be less reliably used. Long conversations can lead to drift in understanding as earlier context fades.

What AI Coding Assistants Actually See

When you paste code into an AI coding assistant, the model doesn't see characters or lines in the same way you or I do. It sees **tokens**: chunks of text that might be whole words, parts of words, or individual characters depending on how common they are in the model's training data. The word "function" is probably a single token. An unusual variable name might be split into several tokens.

This matters because AI models have "context windows"–limits on how many tokens they can consider at one time. Early models like the original Codex had context windows of a few thousand tokens, enough for a single file but not much more. By 2024, context windows had expanded dramatically. Claude's window grew to 200,000 tokens, and later versions approached one million. GPT-4 and other models followed similar trajectories.

To put this in perspective: 200,000 tokens is roughly about 150,000 words, or about 500 pages of text–more than all of the pages in this book. A million tokens approaches the length of a substantial novel. This sounds like a lot, but legacy codebases are very large. A single COBOL application might span hundreds of programs, copybooks, and JCL scripts. The Social Security Administration's systems contain 60 million lines of COBOL. The Context window for current tools can't accommodate that all at once. At least not yet.

This creates practical constraints. When analyzing legacy systems, you can't simply feed the entire codebase to the model and ask for an explanation. You have to work in chunks, providing relevant context for each analysis. The model's understanding of any particular piece of code is shaped by what else fits in the window alongside it. Information at the edges of the context may be less reliably used than information in the middle.

The expanding context windows represent genuine progress. Tasks that were impossible in 2021 are routine today. But understanding the constraint helps explain both why these tools work as well as they do and why they sometimes miss connections that span beyond their window of attention.

The verification problem is perhaps the most fundamental limitation. AI-generated code still requires human review. If you can't verify the output, you can't trust it. This shifts where expertise is needed, but it doesn't eliminate the need for expertise. As one commenter on a discussion about AI and COBOL modernization put it: "Any idiot can do a good job of taking 1 COBOL program and convert it to something else. But putting these multiple converted programs together will always be a big fail."[11] The challenge isn't translating individual pieces of old code into more modern code. It's ensuring the whole system works correctly when it's put together.

What AI can't do includes understanding organizational context it hasn't been given, making judgment calls about business priorities, verifying its own output for correctness, replacing domain expertise, and guaranteeing that generated code is correct. These limitations aren't temporary gaps that will be filled by the next model release. They reflect fundamental aspects of how these tools work.

The appropriate stance is to treat AI output as a best first draft requiring expert verification, not as the final truth. Use AI for what it's good at, and humans for what they're good at. Build processes that assume AI will make some mistakes. Design verification to catch errors before they reach production.

Why This Matters for Legacy Systems

AI coding assistants are particularly valuable for legacy system modernization because they address specific challenges that have made modernization so difficult.

The language barrier is perhaps the most obvious. Legacy systems use languages that few people know anymore: COBOL, Assembly, RPG, PL/I, Natural. New developers actively avoid learning these languages. The knowledge pool shrinks every year as experienced practitioners retire (and, as discussed in Chapter 1, expire). AI tools can read and explain these languages even when developers can't, because the models were trained on legacy code that exists in their datasets.

Consider what this means in practice. A developer who has never written a line of COBOL can paste an unfamiliar program into an AI assistant and receive an explanation of what it does. Not a translation to another language, but a plain-English description of the business logic it has been given. The developer can ask follow-up questions, request clarification on specific sections, and build understanding incrementally. This was essentially impossible before these tools existed.

Julia Kordick, who works on COBOL modernization at Microsoft, describes this dynamic directly. She modernizes COBOL systems without ever having learned COBOL herself. As she puts it: "I didn't need to become a COBOL expert. Instead, I focused on what I do best: designing intelligent solutions. The COBOL experts provided the legacy system knowledge."[12] The AI handles the language barrier; humans provide the domain expertise and judgment.

The documentation gap is another challenge these tools address. Legacy systems are notoriously under-documented. Documentation that exists is often outdated or wrong. The people who actually understood the system may already be gone. AI can generate documentation by analyzing the code itself, creating artifacts that can be verified even by people who can't read the underlying code.

The scale problem has always made legacy modernization daunting. These systems often contain millions of lines of code. Manual analysis at this scale would take years. AI can process large codebases relatively quickly. Patterns that would take humans months to find can be identified in hours. This changes the underlying economics of modernization, making projects feasible that would previously have been impractical.

There's also a knowledge preservation opportunity that these tools enable. The traditional approach to modernization is to translate old code into new code and hope it works. An AI-enabled approach can extract understanding first, and then form the basis of a translation to new code. The extracted understanding is itself valuable: documentation that outlasts the current system and serves as a foundation for future work.

The Unique Challenges of Legacy Code

While AI tools offer significant advantages for legacy modernization, legacy systems also present unique challenges that generic AI tools aren't designed to handle.

Legacy code is simply different from modern code. It was written in an era with different constraints and idioms.

Mainframe patterns don't match modern web development. Batch processing dominates rather than real-time interaction. Data lives in files rather than databases. The eighty-column card format imposed constraints that persist in code structure decades later. Variable names are often cryptic, limited by conventions from an era when memory and storage were precious.

Institutional context matters enormously and is almost never in the code. Why does this calculation round down instead of up? Why is this edge case handled this way? What policy change from years past explains this conditional logic branch? An AI tool won't know organizational history because the answers are not in the code itself. They are in the heads of people who may be moving close to retirement, or in policy documents that were never linked to the technical implementation.

This is the "why" problem. AI can often tell you what code does. It often can't tell you why. The policy rationale behind code written thirty years ago isn't documented in the source. This is where human expertise remains essential, and where the retiring workforce crisis described in Part I becomes acute.

The integration challenge compounds the difficulty. Legacy systems don't exist in isolation. They have dependencies on other legacy systems, complex batch job schedules, data flows across multiple components. AI analyzing one component can miss the bigger picture. The interactions between systems are often where the real complexity in a technology system lives.

Customization and training can improve results significantly. Generic prompting of AI tools can sometimes produce mixed results with legacy languages. Early experience has shown that without specific guidance about COBOL patterns, idioms, and domain context, output quality can vary widely. Custom instruction sets that teach the model about specific patterns and terminology produce much better results. This is an evolving area that we'll discuss more about in the next chapter.

The verification gap presents a particular challenge for legacy modernization. Who verifies that AI understood the legacy code correctly? If no one on the team can read COBOL, how do you know the AI got it right? This is where the specification-driven approach that the next chapter explores becomes essential. The answer is to generate artifacts that non-programmers can verify, rather than code that only programmers can evaluate.

A Different Division of Labor

The emergence of AI coding assistants suggests a new model for legacy modernization, one that plays to the strengths of both humans and AI tools.

AI is good at reading and parsing code at scale. It excels at identifying patterns and dependencies. It can generate documentation and explanations tirelessly. It translates between representations without fatigue. It's consistent and it's fast.

Humans are good at understanding organizational context. We make judgment calls about priorities. We verify business logic correctness. We recognize when something doesn't make sense even if we can't articulate exactly why. We know the "why" behind the "what."

The new model for legacy modernization introduced in Section III pairs these capabilities. AI analyzes legacy code and generates explanations. Humans verify that these explanations are correct. AI (in concert with human engineers) generates modern implementations. Humans verify that these implementations are correct. Each party does what they do best.

This partnership changes what skills a legacy system modernization team needs. The need for rare COBOL experts who can read and modify legacy code directly is reduced (which is good because there are fewer and fewer of them around). There's more need for domain experts who understand the business logic behind system design and can verify that generated specifications are correct. There's more need for engineers skilled with AI tools who can guide the analysis and generation process effectively.

The skill mix changes, but skill remains essential. AI hasn't made expertise obsolete. It has changed what expertise matters most. For legacy modernization specifically, domain knowledge becomes more valuable relative to legacy language expertise. The ability to explain clearly, verify carefully, and think systematically about complex systems matters more than ever.

What Comes Next?

AI coding assistants represent a genuine capability shift, not just incremental improvement. Their limitations are real and must be respected. But for legacy modernization specifically, they address some of the hardest problems: the language barrier, the documentation gap, the scale challenge, the knowledge preservation imperative.

The question that remains is how to use them most effectively. Direct code translation is fast but can be dangerous. Errors in understanding can get embedded in newly generated code. Domain experts can't verify technical implementations if they can't read code. Institutional knowledge remains locked away, now in a different, more modern language but with the same opacity.

What if we used AI's analytical capabilities differently? What if the goal wasn't translating old code into new code but capturing knowledge? What if AI generated artifacts that non-programmers could verify, rather than code that only programmers can evaluate? This is where the specification-driven approach enters.

The next chapter explores how AI makes comprehensive software specifications practical, and why treating specifications as the primary artifact, rather than code, changes the legacy modernization equation fundamentally. We've established what AI can do. The next question is what AI *should* do. The answer, taken up in the following chapters, lies in using AI to generate specifications that humans can verify, rather than code that humans can't.

Chapter 5: Specification-Driven Development

As I started working with AI coding assistants, I kept running into the same problem: momentum between sessions. I would make real progress on a project, then step away for a few days or a few weeks. When I returned, the AI had no memory of where we had been. I found myself spending the first hour of each session trying to reconstruct context that had existed naturally the session before.

So I started experimenting. At the end of each session, I would ask my AI assistant to summarize what we had done, the decisions we had made, and where we should pick up next time. I kept these summaries. And when I started a new session, I would paste them into the initial prompt along with other context: the goals of the project, the constraints we were working within, the patterns we had agreed to follow.

My initial prompts kept getting longer. I was writing paragraphs of context before asking the AI to do anything. This felt strange, even wrong. Everything I had absorbed about prompt engineering suggested brevity. Yet the longer, more contextual prompts produced dramatically better results for me. The AI wasn't just completing code; it was working within a framework I had established. It understood not just what I wanted done, but why. I felt like I was bending the rules. Maybe even breaking them.

Then a colleague shared a YouTube video of a talk that reframed everything I had been experiencing. Sean Grove, who works on alignment research at OpenAI, gave a presentation called "The New Code" at an AI engineering conference. In it, he articulated something I had been feeling but couldn't put a name to.[1]

Grove pointed out something that seems obvious once you hear it: when we use AI to generate code, we typically keep the code and throw away the prompt. We treat the generated artifact as the valuable output and discard the instructions that produced it. But this is backwards. He compared it to carefully version-controlling a compiled binary while shredding the source code.

"We communicate via prompts to the model," Grove said, "and we tell them our intentions and our values and we get a code artifact out at the end and then we sort of throw our prompts away. They're ephemeral." He noted that when you compile TypeScript or Rust, "no one is happy with that binary. That wasn't the purpose. It's useful. In fact, we always regenerate the binaries from scratch every time we compile... It's the source specification that's the valuable artifact. And yet when we prompt models, we sort of do the opposite. We keep the generated code and we delete the prompt. And this feels like you shred the source and then you very carefully version control the binary."[2]

The prompt, Grove argued, is the source. The code is the compiled output. If the specification contains enough information to generate the code, then the specification is the more valuable artifact. You can always regenerate the code from the specification.

This clicked for me. Those long initial prompts I had been writing weren't a workaround. They weren't rule-breaking. They were proto-specifications. I had started to practice specification-driven development without knowing it had a name.[3]

The Historical Problem with Specifications

Specifications have always been recognized as valuable artifacts in software development in principle. The challenge has always been that writing comprehensive specifications took enormous effort. They required skills and time that projects often did not have. When they were created, they tended to drift out of sync with code almost immediately. The result: a pattern where documentation was either skipped entirely, generated only rudimentarily, or generated once and then never maintained.

Traditional waterfall approaches to software development try to front-load specifications. But those specifications are often based on imagined requirements, not working systems. They capture what stakeholders thought they wanted, not what systems actually did. When applied to legacy modernization, this created a fundamental mismatch. You can't specify what you want from a new system without understanding what the old system does. And understanding what the old system does is exactly what legacy modernization struggles with.

Agile approaches emerged partly as a reaction to the front-loaded, specification-heavy waterfall model. "Working software over comprehensive documentation" became the mantra. The primary measure of progress is a functional product, not a detailed document; documentation is incidental to working code. Agile advocates for "just enough" and "just-in-time" documentation, creating it only when needed, at the last responsible moment, and keeping it lean to avoid waste. In practice, this means documentation can be overlooked or neglected, particularly by teams new to agile or not yet mature in their practice of it—a description that fits many agile teams in government.

Agile does include the concept of automating documentation where possible to reduce the pain of writing and maintaining documentation. This automation principle aligns with the SpecOps approach of using AI tools to help generate comprehensive system specifications. But even with automation, agile's emphasis on minimal documentation creates gaps when applied to legacy modernization.

Both waterfall and agile approaches fail at the same thing: creating and maintaining comprehensive documentation of what systems actually do. Waterfall tries to specify before building and gets specifications that don't match reality. Agile tries to document after building and gets documentation that is incomplete or completely absent. Neither solves the problem.

What AI Changes

AI can analyze existing code and generate specifications from it. This reverses the traditional sequence: the specification emerges from the system, not before it. AI can draft specifications faster than humans can, making comprehensive coverage more practical. The human role shifts from writing specifications to verifying them. Verification is faster and easier than creation from scratch. The bottleneck moves from "can we create specifications?" to "how quickly can we verify them?"

An AI tool can analyze thousands of lines of code in minutes. It can identify patterns, trace data flows, and map dependencies. This doesn't mean it understands the code perfectly. Chapter 4 discussed the limitations of AI tools. But it does mean that the mechanical work of reading and summarizing code can be accelerated dramatically. A human analyst might spend days mapping a single module; AI can draft that analysis in minutes, or hours.

AI's natural language capabilities allow it to express technical concepts in accessible terms. Code that only COBOL experts could read becomes descriptions that policy experts can understand. This translation from technical to accessible is exactly what specifications need to be useful. The specification becomes a bridge between technical implementation and business understanding.

Modern AI tools can hold an increasingly large amount of context (a feature that is improving at breakneck speed). This means specifications for related components can be developed with awareness of each other. Cross-references, dependencies, and interactions can be captured. The specification becomes a coherent description of the system, not just isolated component documents.

Creating specifications used to be the bottleneck. Now verification is the bottleneck. This is actually a better problem to have. Verification can be done by domain experts who understand policy and organizational requirements. It doesn't require the rare technical expertise needed to read legacy code. The pool of people who can participate in quality assurance expands dramatically.

Specifications as Source Code

Grove went beyond just reframing prompts as specifications. He argued that code itself is a "lossy projection" from a specification. Just as decompiling a C binary doesn't give you nice comments and well-named variables, code doesn't fully embody the intentions and values that produced it. You have to infer what the team was trying to achieve. The specification captures what's lost in translation to code.[4]

This means the specification contains more information than the code. A robust specification can generate not just code, but documentation, tutorials, test cases, and more. Grove noted that "a sufficiently robust specification given to [AI] models will produce good TypeScript, good Rust, servers, clients, documentation, tutorials, blog posts, and even podcasts."[5] The specification is the source from which all these outputs can be derived. The code is just one possible output of many.

Flipping the Script on Documentation

Aspect	Traditional Pattern	Specification-Driven Pattern
Sequence	Design → Code → Document (hopefully)	Analyze → Specify → Verify → Implement
Source of Truth	Code is authoritative; documentation eventually follows	Specification is authoritative; code always follows
Drift	Documentation drifts out of sync	Code changes require specification changes first
Knowledge Location	Knowledge lives in code	Knowledge lives in specifications
Who Can Verify	Only those who read the code	Domain experts who understand policy and organizational needs

Traditional development treats this relationship backwards. Design leads to code, which might eventually produce documentation. Documentation is derivative of code. When code changes, documentation should change, but rarely does. The result: documentation that's incomplete, outdated, or both. For government systems, this pattern has played out over decades.

Specification-driven development inverts the relationship. The specification is authoritative; code is derivative. When behavior needs to change, the specification changes first. Implementation follows specification, not the other way around. This mirrors how GitOps treats infrastructure: the declared state is authoritative, the actual state follows.[6] More on this in the next chapter.

For legacy modernization, this inversion matters a lot. Instead of translating old code into new code, we aim to extract understanding into specifications. Verify that understanding with domain experts. Then we can generate new implementations from verified specifications. The specification becomes the persistent artifact. Code implementations can come and go; the specification endures.

This approach eliminates the "we'll document it later" problem that has plagued software development for decades. Documentation isn't created after the fact; it's the starting point. The specification exists before the new code is written. There's no gap to close. For the first time, comprehensive documentation becomes a natural byproduct of the development process rather than an afterthought that sometimes doesn't happen.

Teaching AI Your Context

Getting useful specifications from AI requires teaching it your specific context. AI tools come with general knowledge of programming languages and patterns. But they don't know your organization's specific systems, terminology, or business domain. A generic COBOL analysis may miss organization-specific conventions. Domain-specific business rules will be lost without context. The quality of AI output depends heavily on the quality of instructions provided.

What Does a Good System Specification Look Like?

Characteristic	Description
Behavior-focused	Describes what the system does, not how it's implemented
Accessible language	Uses terms domain experts understand, not just developers
Concrete examples	Includes specific examples for complex rules and calculations
Edge cases documented	Captures exceptions, boundary conditions, and special handling
Uncertainties noted	Flags areas requiring additional verification or further investigation
Source references	Points to authoritative sources like policy documents or regulations
Verifiable by non-coders	Can be reviewed by someone who doesn't read the implementation language

Instruction sets, sometimes called "skills" or "custom instructions," guide AI behavior. They provide important context about legacy platforms, business domains, and specification formats. They teach AI about your specific patterns and conventions. They establish terminology and glossary. They define what "good" specifications look like for your context.

These instruction sets fall into several categories. Legacy code comprehension instructions help AI understand COBOL, mainframe environments, and legacy databases. Specification generation instructions guide consistent structure, plain language, and edge case identification. Domain-specific instructions provide context for benefits eligibility, tax calculations, or regulatory compliance. Each category serves different needs in the specification process.[7]

It's also worth noting that instruction sets can be shared across organizations. A well-crafted COBOL comprehension skill works whether you're modernizing a benefits system in California, a tax system in New York, or a licensing system in Texas. The legacy language patterns are similar even if business domains differ. Government agencies don't need to start from scratch. This creates opportunity for collaboration at a level that code sharing rarely achieves.[8] Instruction set development happens in parallel with specification work. Each specification project teaches you what instructions work and what's missing. The instruction library grows over time, becoming more valuable with each project.[9]

Why Specifications Outlast Code

Code is always tied to specific technologies: languages, frameworks, platforms, architectures. When those technologies become obsolete, the code is considered "legacy." This is how we got into the current crisis: systems built on technologies that are now decades old. Modern code written today will face the same fate eventually. The technology choices we make now create constraints for those that follow us.

Specifications transcend implementation. A well-written specification describes behavior without specifying how it's implemented. "The system determines eligibility based on income below 150% of the federal poverty level" doesn't depend on COBOL, Java, Ruby, or any other programming language. The same specification can guide implementations on mainframes, cloud platforms, or technologies that don't exist yet. The specification captures the essential knowledge; code is just one possible expression of it.[10]

Government systems encode decades of policy decisions, edge cases, and accumulated understanding. This knowledge currently lives in code that fewer and fewer people can read and understand. When extracted into specifications, it becomes accessible to anyone who understands the domain. Policy experts can read specifications; they can't read COBOL. The specification preserves institutional knowledge in a form that can be maintained and transferred.

In twenty years, the code we write today will need modernization. The languages and frameworks we use will be dated or obsolete. But well-written specifications will still describe what the system does. They'll provide a foundation for the next modernization. We break the cycle of periodic crises by preserving knowledge in a more durable form.

Specification Writing as a Universal Skill

Grove pointed out that specifications aren't unique to programming. Product managers write product specifications. Lawmakers write legal specifications. The US Constitution, he noted, is literally a national model specification: written text that's aspirationally clear and unambiguous, with a versioned way to make amendments, judicial review to assess alignment, and precedent that reinforces the original policy.[11] Everyone who communicates intent and values clearly is writing specifications. The skill transfers across domains.

Grove asked his audience how many felt that code is the most valuable artifact they produce. His answer: code is only 10 to 20 percent of a programmer's value. The other 80 to 90 percent is structured communication. Talking to users, understanding challenges, distilling requirements, planning solutions, sharing plans, testing outcomes. These are all specification activities. "Structured communication is the bottleneck," he said. "And the more advanced AI models get, the more we are all going to starkly feel this bottleneck. Because in the near future, the person who communicates most effectively is the most valuable programmer."[12]

This resonated with everything I had learned working in government technology. The challenge was never primarily about writing code. It was about understanding what needed to be built, why it needed to work a certain way, and whether the thing you built actually achieved its purpose. These are communication problems. These are, fundamentally, specification problems.

Legacy modernization has always required more communication than greenfield coding. Understanding what systems do, why they work the way they do, what they should do instead. These are the hard problems; writing code is comparatively straightforward. Specification-driven development recognizes this and makes specification the explicit deliverable. It aligns the work with what was always the actual challenge.

"Whoever writes the spec," Grove concluded, "be it a PM, a lawmaker, an engineer, a marketer, is now the programmer."[13] This expands who can contribute to system development. Domain experts who understand the business can write or verify specifications. They don't need to code; they need to communicate clearly. For government, this means policy experts can participate in ways they never could before.

Where This Leads Us Next

AI coding assistants represent a genuine capability shift, but the shift isn't just about faster code generation. The more profound change is that AI makes comprehensive specifications for monolithic legacy systems practical for the first time. This enables an inversion: specifications become the source of truth, with code as an implementation of that spec.

For legacy modernization, this addresses the knowledge preservation challenge that traditional approaches typically ignore. The specification captures what systems do in a form that domain experts can verify, that persists regardless of implementation technology, that provides a foundation for future modernization, and that represents institutional knowledge in accessible form.

The instruction sets that guide AI can be shared across organizations, multiplying the benefit. A library of instruction sets for COBOL comprehension, benefits eligibility rules, tax calculations, and other common government domains could accelerate modernization across agencies without each one starting from scratch.

This approach draws on a proven pattern: treating version-controlled specifications as the authoritative source of truth, just as GitOps does for infrastructure. The next chapter examines this parallel in more depth. GitOps transformed infrastructure management by treating configuration files as authoritative.

What can that teach us about applying the same principle to system behavior? The answers illustrate both why this approach works and how we can implement it.

Chapter 6: The Infrastructure as Code Parallel

In 2017, a company called Weaveworks experienced what should have been a catastrophic infrastructure failure. Their entire Kubernetes cluster went down. In a previous era, this would have meant days or weeks of painful reconstruction: manually rebuilding servers, reconfiguring networks, reinstalling software, hoping the existing documentation was accurate enough to get things working again. Instead, they recovered in hours.

Their infrastructure was defined entirely in configuration files stored in the version control system Git. When the Kubernetes cluster failed, they simply pointed their automation tools at those files and let the system reconstruct itself. Everything came back exactly as it had been, because "exactly as it had been" was defined in version-controlled specifications. The running infrastructure in their Kubernetes cluster was an implementation of what was contained in the specification.

When Alexis Richardson, Weaveworks' CEO, told the story of this recovery, people kept asking: "What did you need to have in place to make that work?" The answer—it turned out—was deceptively simple. Everything had to be described declaratively. Everything had to live in version control. And there had to be tools that could continuously reconcile the actual state (the running infrastructure) with the declared state (the version-controlled specification).[1]

This wasn't magic. It was the result of a disciplined approach that the technology industry had been developing for over a decade: Infrastructure as Code (IaC). And the specific practice Weaveworks used, which they called GitOps, represented a mature form of that approach. The Git repository containing the

IaC was the source of truth. The running infrastructure was a derivative artifact, regenerated from those specifications whenever needed.

The pattern solved problems that had plagued operations teams for decades: configuration drift, undocumented changes, inconsistent environments, and the brittleness of infrastructure that no one fully understood. The solution was to stop treating running infrastructure as the source of truth and start treating declarations about infrastructure as the source of truth. Which brings us to legacy systems.

The problems that IaC solved for infrastructure are the same problems we face with legacy code. Systems drift from their documentation. Changes accumulate without clear records. Environments become inconsistent. And, over time, the running system becomes a mystery that no one fully understands.

For infrastructure, the solution was IaC. For legacy systems, the solution is specifications. The pattern is the same: stop treating the running artifact as authoritative and start treating the description of what the artifact should be as authoritative.

The Problem Infrastructure as Code Solved

Before automation changed the landscape, provisioning and configuring infrastructure was a manual, error-prone process. System administrators connected to servers individually, ran shell commands, installed software, modified configuration files, and hoped they remembered to do the same things in the same order on every server. The documentation, if it existed at all, quickly drifted out of sync with reality.

This manual approach created several interrelated problems. Configuration drift was perhaps the most insidious. Small changes to configurations accumulated over time. Someone patched one server but not another. A configuration file was modified to fix a bug and never documented. Environments that were supposed to be identical diverged in subtle ways that caused mysterious failures.

The knowledge problem compounded the drift problem. Understanding what a server actually did required examining the running system itself. The documentation described what someone intended at some point in the past; the running system was the only authoritative source of information on the current state. But servers aren't self-documenting. Configuration can be scattered across dozens of files in different formats. Understanding the actual state requires expertise and time.[2]

The first attempts to address these problems came from configuration management tools like CFEngine, Puppet, and Chef. These tools allowed administrators to define desired configurations in code rather than applying them manually. Run the same script on multiple servers, and they would converge to the same state.

This was a genuine improvement, but it didn't fully solve the problem. Early configuration management tools often mixed imperative and declarative approaches. The running infrastructure was still the source of truth; the configuration management scripts were just a way of managing that source of truth more reliably.

The conceptual breakthrough came with fully declarative tools and a more disciplined use of version control. Instead of writing scripts that modified infrastructure, you wrote declarations of what infrastructure should exist. The tooling figured out how to get there.[3]

This shift had profound implications. When infrastructure is declared, not scripted, the declaration is readable by humans who understand the domain even if they don't know the implementation details. When declarations live in version control, you have a complete history of every change, who made it, when they made it, and why. When the declaration is authoritative, discrepancies between the declaration and reality indicate a problem to be fixed, not an update to be documented.

The game changer was treating the declaration as the source of truth. The running infrastructure became a derived artifact, something generated from specifications rather than something that documentation describes.

How GitOps Made It Work

The term "GitOps" was coined by Alexis Richardson in a 2017 blog post. It wasn't a new invention so much as a formalization of practices that had been evolving in the cloud-native community. What made GitOps distinctive was its clarity about what served as the source of truth: the Git repository housing the IaC.[4]

As GitOps matured, the community converged on four core principles. First, the entire system is described declaratively. You don't write scripts that create infrastructure; you write declarations of what infrastructure should exist. This makes the specifications readable and auditable.

Second, the desired system state is versioned in Git. Git provides a complete history of every change. You can see who changed what, when, and why. You can compare the current state to any previous state. You can roll back to a known-good configuration quickly.

Third, approved changes are automatically applied to the system. When you merge a pull request to an IaC repo, the change takes effect. This eliminates manual deployment steps and ensures that what's declared is what's deployed.

Fourth, software agents ensure correctness and alert on divergence. Tools continuously compare the actual state to the declared state. If they drift apart, the system either automatically corrects the drift or alerts humans to investigate.[5]

The Four Principles of GitOps

Principle	Description
Declarative	The entire system is described declaratively, specifying the desired end state rather than the steps to achieve it
Versioned	The desired system state is version controlled in Git, providing complete version history and enabling rollback if needed
Automated	Approved changes are automatically applied to the system, eliminating manual deployment steps
Reconciled	Software agents continuously ensure correctness and alert on divergence between declared and actual state

These principles reinforce each other. Declarative specifications make version control meaningful. Version control makes automated deployment safe. Continuous reconciliation makes the source of truth actually authoritative.

GitOps succeeded because it addressed real operational pain points with a coherent and applicable model. The model was understandable: the Git repo is the source of truth; change the repo, and the system changes; what you see in Git is what you get in production. The model scaled: whether you were managing one Kubernetes cluster or a dozen, the approach was the same. The model was safe: changes went through code review, history was preserved, rollback was trivial.

And perhaps most importantly, the model enabled recovery. When disasters happened, when configurations were corrupted, when someone made a mistake, the response was straightforward: restore from the declarations in Git. The infrastructure wasn't really lost because the true infrastructure lived in a declarative specification. The running servers were just one instantiation of those declarations.

Declarative vs. Imperative: Why It Matters

The distinction between declarative and imperative approaches is fundamental to understanding why specification-driven modernization holds so much promise.

An imperative script to set up a web server might say: "Install package X. Create directory Y. Copy file Z. Start service W." Each step must execute in order. If something changes, you need to figure out which steps to re-run. The script encodes the path from some starting point to a destination, but that path only works from that starting point.

A declarative specification might say: "A web server should exist with package X installed, directory Y present, file Z in place, and service W running." The declaration doesn't specify how to achieve this. A tool reads the declaration, examines the current state, and figures out what changes are needed. If the package is already installed, don't install it again. If the service is running, leave it alone.[6]

This distinction matters because systems change over time. The imperative script assumes a particular starting state. The declarative specification accommodates any starting state. As configurations drift, as components are added or removed, the declarative approach remains applicable while the imperative approach becomes increasingly fragile.

Configuration drift is the gradual divergence of a system from its documented or intended state. In an imperative world, drift is nearly inevitable. Someone makes a manual change. An automated process modifies something. A random dependency gets updated. Each small change is individually reasonable, but the cumulative effect is a system that no longer matches its description.

Declarative approaches handle drift differently. The specification describes the intended state. Tools continuously compare actual state to intended state. Deviations are detected and either automatically corrected or flagged for human review. The specification remains authoritative because the tooling enforces it.

For legacy systems, drift is an even bigger problem than for infrastructure. Business logic changes accumulate over decades. Workarounds become permanent. The code that runs today bears little resemblance to what was originally documented. Understanding the actual behavior of the system requires examining the actual code running in production because no up to date specification has been maintained.

The Parallel to System Specifications

An infrastructure declaration describes what infrastructure should exist: which servers, with what configurations, connected in what ways. A system specification describes what behavior should exist: which inputs produce which outputs, under what conditions, with what constraints.

Both are descriptions of intended state. Both can be version-controlled. Both can be verified against reality. Both can guide implementation. The parallel is direct, even if the details differ.

In GitOps, you might declare: "A load balancer should exist that routes traffic to three application servers based on health checks and round-robin distribution." This declaration is readable, verifiable, and implementation-agnostic.

In SpecOps, you might specify: "An eligibility determination should accept applicant data and return a determination based on income thresholds, household size, and qualifying exemptions as defined in Section 4.2 of the program regulations." This specification is readable, verifiable, and implementation-agnostic. It could be implemented in different technology stacks, on different platforms, with different architectures.[7]

The connection between declaration and implementation works the same way. In GitOps, tools read infrastructure declarations and provision corresponding resources. In SpecOps, developers read system specifications and generate corresponding code—whether that's AI generating code autonomously, AI assisting human developers, or human developers working entirely on their own. In all cases, the specification guides the implementation.

SpecOps Similarity to GitOps

GitOps	SpecOps
Infrastructure declarations	System specifications
Git repository	Specification repository
Terraform, Kubernetes manifests	Specification documents
Continuous reconciliation	Domain expert verification
Automated deployment	AI-assisted code generation
Infrastructure drift detection	Specification conformance testing

One of GitOps' greatest strengths is the history that version control provides. Every change is recorded. You can trace who made what change, when, and why. System specifications in a Git repository gain these same benefits. When specifications evolve, the evolution is recorded. When business rules change, the change is documented in the commit. When someone asks "why does the system work this way?", you can trace back through the history to find when that behavior was specified and why.[8]

For legacy systems, this detailed history often doesn't exist. The code has likely evolved through hundreds or thousands of changes over time, but no single source captures what changed and why. Creating specifications for legacy systems and managing them through version control establishes the history going forward, even if the past is murky.

In GitOps, tools continuously reconcile actual infrastructure state with declared state. For system specifications, the analogous process is verification. After generating code from specifications, you test whether the code actually implements the specified behavior. If tests fail, either the code is wrong or the specification is wrong. Either way, the mismatch is visible.

An important difference is that infrastructure reconciliation can often be automated, while system verification requires human judgment. Determining whether code correctly implements business logic requires domain expertise. This is why SpecOps emphasizes domain expert verification; it's the human reconciliation that ensures specifications are accurate. More on this in the following chapters.

Lessons from the GitOps Model

The evolution of GitOps offers practical guidance for legacy modernization. The most important lesson is that declaring something as the source of truth isn't enough. The source of truth must be enforced. In GitOps, this enforcement comes from continuous reconciliation: tools that constantly check actual state against declared state and correct discrepancies.

For specifications, enforcement comes from process discipline. All changes must flow through specifications first.

Code generation derives from specifications. Testing verifies against specifications. When shortcuts are taken, when code is modified without updating specifications, the source of truth erodes.[9]

Organizations didn't adopt GitOps all at once. They started with what they had. They automated one thing, then another. They moved toward declarative management gradually. The full benefits came later, but value was delivered at every step.

For legacy modernization, the same approach applies. You don't need to specify the entire system before gaining benefits. Specify one component. Verify it. Generate modern code. Learn from the process. Then repeat. Each component specified is knowledge captured, regardless of whether other components follow immediately.

GitOps adoption also changed how teams worked together. Developers could see infrastructure configurations. Operations could review proposed changes before they took effect. Everyone had visibility into the state of systems. The audit trail satisfied compliance requirements. The clarity streamlined onboarding of new staff.

Specification-driven modernization creates similar organizational benefits. Domain experts can review specifications even if they can't review code. New team members can understand systems by reading specifications. Auditors can examine specifications for compliance. The knowledge is accessible, not locked in code that requires deep technical expertise to interpret.[10]

How GitOps Changed the Game

For decades, the running system was authoritative. Documentation described what the system was supposed to do; the system itself determined what it actually did. When they diverged, everyone understood that the documentation was wrong, or at best outdated. You couldn't trust documentation. You had to trust the running system. There was no alternative.

GitOps reversed this. The declarations in Git are authoritative. The running infrastructure should match. When they diverge, the infrastructure is wrong, or at least, something requires attention. You trust the declarations. The infrastructure is just their current implementation.

Applying this inversion to legacy systems is what makes SpecOps different from traditional modernization. Traditional approaches treat the legacy code as authoritative. You study it, translate it, verify that the translation matches. The code is the truth.

SpecOps proposes an inversion. The specification becomes authoritative. The legacy code is evidence to be analyzed. The modern code is an implementation to be generated. Both the old and new code should match the specification. The specification is the truth.

This has practical consequences. When legacy code is authoritative, errors in the code are features. A bug that's been running for twenty years is "correct behavior" because that's what the system does. Translating it faithfully means replicating the bug.

When the specification is authoritative, errors in the code are errors. The specification describes what the system should do. If the code diverges, that's a discrepancy to resolve. Maybe the specification is wrong and needs updating–that is a question domain experts can have input on. Maybe the code is wrong and the new implementation should fix it. Either way, the discrepancy is visible and requires a decision.

When legacy code is authoritative, verification is impossible for non-technical experts. They can't read the code. They can only hope the technical experts understood the governing policy correctly.

When the specification is authoritative, verification is possible for domain experts. They can read the specification. They can confirm it describes the intended behavior. They can participate meaningfully in ensuring the correctness of a system.

When the specification is authoritative, the value of modernization includes the specification itself. Even if you never generate modern code, the verified specification can have tremendous value. It documents what the system does. It aggregates institutional knowledge. It serves purposes beyond just a new implementation.

This inversion is the conceptual foundation on which the SpecOps methodology rests. Everything that follows in Part III builds on this idea. The six phases of SpecOps, the emphasis on domain expert verification, the practice of having changes flow through specifications, the expectation that specifications outlast implementations: all of it assumes that specifications are the source of truth.

The previous chapter argued for this on first principles: specifications can be verified by domain experts, specifications are more portable than code, specifications capture knowledge in accessible form. This chapter has provided supporting evidence from practice: the same approach has transformed technology infrastructure management.

Part III will introduce the SpecOps methodology itself. It will explain the compilation vs. transpilation distinction that frames the approach. It will walk through the phases from Discovery through Deployment. It will detail the core principles and how they work together.

The infrastructure world learned painful lessons about managing complex systems over many years. Those lessons matured into a clear pattern: declare the intended state of infrastructure, version control those declarations, continuously reconcile actual state with intended state. The pattern worked because it made the source of truth explicit, visible, and enforceable.

Legacy systems face the same underlying challenges: drift from intent, undocumented changes, knowledge locked in running artifacts. The same pattern can address those challenges. The tools are different, the artifacts are different, but the conceptual approach is the same.

SpecOps is, in essence, GitOps for system behavior. The pattern has been proven in infrastructure. Now it's time to apply it to legacy systems.

Part III

Introducing SpecOps

"If there be nothing new,
but that which is
Hath been before,
how are our brains beguiled
Which, laboring for invention, bear
amiss"

Sonnet 59

Chapter 7: Compilation vs. Transpilation

In the summer of 2024, I worked with a colleague on a coding challenge that changed how I think about legacy system modernization. The Beeck Center for Social Impact + Innovation at Georgetown University had organized the Policy2Code Prototyping Challenge, inviting teams to explore innovative ways to use generative AI to improve public benefit delivery. Our team took a different approach than most.[1]

While many teams focused on enhancing applicant experiences or assisting benefit navigators, we focused on a friction point we'd observed repeatedly in public benefit systems: the gap between the people who are experts in complex program policies and those who are experts in software implementation. This lack of shared understanding makes new system implementation difficult, costly, and error-prone. Policy experts can't verify code. Engineers don't fully grasp policy nuance. The result is systems that don't quite work the way they should.

Our approach was to use large language models to generate an intermediate format between policy language and software code. Not policy directly to code, but policy to a domain-specific language that both policy experts and software engineers could understand.[2] This intermediate representation empowered policy experts to verify correctness before any code was written. When we demonstrated this work at the BenCon 2024 conference, our team was recognized with an award for Outstanding Policy Impact.

That experience crystallized something my colleague and I had been thinking about for a while. The pattern that worked wasn't "use AI to convert policy rules directly into code." The pattern that worked was: use AI to generate an intermediate format that humans could verify, then use that verified intermediate format to generate code. Using the verification step in the middle was our aha moment.

After more than two decades working in government technology, spending a non-trivial amount of that time modernizing legacy systems, I came to the uncomfortable conclusion that I had probably been doing it wrong the whole time.[3] The traditional approach to legacy modernization focuses almost exclusively on one thing: converting old into new. Old policy into new code. Old code into new code. But the pattern from our Policy2Code work suggested a different approach entirely.

What if the intermediate format wasn't just a step along the way, something to be discarded, but the valuable artifact itself?

The Traditional Modernization Assumption

When we talk about modernizing a legacy system, we almost always mean rewriting it in a modern language or technology platform. The legacy code is the input; modern code is the output. Everything in between, the requirements documents, architecture diagrams, and test plans, exists solely to support that conversion process. If these projects finish successfully, and that's often a big "if," the new code becomes the source of truth for how the system works, and these intermediate documents go away.

This assumption is so deeply embedded that we rarely question it. The goal is producing working modern code. Once we have that, we're done.

AI coding assistants seemed to accelerate this approach dramatically. Feed COBOL into one end, get Java out the other. Early experiments showed this was technically feasible at scale. The promise was tantalizing: automate away the expensive, time-consuming translation work. "Just convert the code" seemed like the obvious way to use AI in legacy modernization.

But consider what direct translation actually produces. You get modern code that *hopefully* behaves like the old code. You get no verified understanding of why the code works the way it does in the first place. Business logic remains embedded in implementation, just in a different language. Errors in the AI's understanding during this transpilation process get baked into the new codebase. And you have a new legacy system, with technical debt accumulating from day one.

The verification problem is particularly acute. When AI translates COBOL to Java, who verifies it's correct? Engineers can review the code, but do they understand the business rules embedded in thirty-year-old logic? The people who understand policy typically can't read Java any better than they can read COBOL. Testing catches some errors, but only the ones your tests anticipate. Problems discovered in production are the most expensive to fix.

And after all that work, the knowledge of how the system is supposed to work remains trapped. It's just locked in a different code base. The tax policy expert still can't verify the implementation. The eligibility determination logic is still opaque to the program administrators who are supposed to understand it. We've moved the furniture around, but we haven't solved the fundamental problem.

The Compiler's Insight

There's a different way to think about what we're trying to accomplish. The distinction between transpilation and compilation offers an alternative mental model.

Transpilation is the process of converting source code written in one programming language directly into functionally equivalent source code in another programming language.[4] The goal is a direct translation: COBOL gets turned into Java, mainframe to cloud-native, legacy to modern. The process is straightforward: input old code, output new code. The ultimate goal is equivalence, making the new code do exactly what the old code does, even if we don't fully understand what that is.

Compilation works differently, and the Java programming language offers a useful illustration. When a developer writes Java code and compiles it, the code doesn't get translated directly into instructions for a specific computer platform. Instead, the Java compiler produces something called bytecode, an intermediate representation that isn't tied to any particular processor. This bytecode runs on the Java Virtual Machine (JVM), a software layer that exists for almost every operating system and hardware platform. The JVM takes the bytecode and translates it into the specific instructions that different machines understand.

This intermediate step is what made Java's key feature of "write once, run anywhere" possible. A developer could compile Java source code into bytecode on a Windows machine, and that same bytecode would run on Linux, Mac, mainframes, or anything else with a JVM installed. The bytecode captures what the program does without being tied to how any specific platform executes it. The meaning of the program is preserved in a platform-independent form.

Transpilation vs. Compilation

TRANSPILATION (CODE-TO-CODE)

Legacy Code (COBOL) → AI Translator → Modern Code (Java)

Direct translation. Knowledge remains locked in code.

COMPILATION (CODE-TO-SPEC-TO-CODE)

Legacy Code (COBOL) → AI Extractor → Specification Human-readable → AI Generator → Modern Code (Java)

Domain Expert Verification
Experts verify the spec before code generation ✓

The specification becomes the source of truth.
Knowledge is preserved, verified, and technology-agnostic.

What if we applied this same concept to legacy modernization? Instead of translating legacy code directly to modern code, what if we first extracted an intermediate representation that captures what the system does, independent of how it's implemented? Instead of code-to-code, what about code-to-specification, and then specification-to-code?

When the specification serves as that intermediate representation, something important shifts. The specification captures the meaning of the legacy system in a form that isn't tied to COBOL or Java or any other language. Like Java bytecode, it's platform-independent. Unlike bytecode, it's human-readable. Domain experts can verify it. And from that verified specification, you can generate implementations in whatever technology makes sense today, or regenerate them in whatever technology makes sense twenty years from now.

This isn't just theoretical. An IT Modernization team from MITRE has been exploring exactly this approach, using LLMs to extract logic from legacy systems. Their research generated and evaluated intermediate representations from legacy code, and their key finding validates the core tenet of SpecOps: LLMs can reliably generate intermediate representations at scale from legacy code.[5]

But MITRE's research revealed something else important. When they tried to use automated metrics to evaluate the quality of AI-generated intermediate representations, those metrics didn't correlate with what human subject-matter experts actually thought was good. The computational measures said one thing; the humans said another. This isn't a limitation of the SpecOps approach. It's confirmation that there's no shortcut around human verification. You can't automate quality assessment for this work. The domain experts have to be in the loop.

MITRE recommends that LLMs be used in a "highly supervised manner" for legacy modernization on mission-critical systems. That's not a warning against using AI. It's an empirical endorsement of exactly what SpecOps prescribes: AI generates, humans verify. The verification step isn't optional overhead. It's essential, and no automated metric can replace it.

This independent research confirms our insight from the Policy2Code challenge. AI can do the heavy lifting of extracting logic and generating intermediate formats. But humans must verify these intermediate artifacts. The intermediate representation creates a checkpoint where that verification can happen by domain experts.

Transpilation vs. Compilation at a Glance

Criteria	Transpilation	Compilation
Primary Output	Modern code	Verified specification + modern code
What's Valuable	Working system	Institutional knowledge preserved
Who Verifies	Engineers (code review, testing)	Domain experts (spec review) + engineers
When Errors Found	Testing or production	Specification review
Long-term Value	Code (temporary)	Specification (enduring)

Making the specification serve as that intermediate representation changes the equation. The specification becomes the thing we verify and preserve. Modern code is then generated from the verified specification. This is compilation in a new sense: gathering together authoritative information about how a system works into a comprehensive, verified document.

The key point is that this approach produces two valuable artifacts instead of one. Transpilation produces modern code. Compilation produces the specification and modern code. The specification captures institutional knowledge in verified, accessible form. Even if the code needs to change again, the specification remains valuable.

With transpilation, the question is: "Can we convert this code to a modern language?" With compilation, it's a fundamentally different question: "Can we understand what this system does, verify that understanding, and then build from that verified knowledge?"

The second question is harder to answer. But answering it produces something much more valuable than just updated code.

What Legacy Code Experts Already Knew

This distinction discussed above between transpilation and compilation, or-maybe more importantly-the distinction between transforming software code and understanding what it does is not new. Software practitioners who have worked intimately with legacy code have advocated for decades that understanding must precede transformation.

Michael Feathers' *Working Effectively with Legacy Code*, published in 2004, remains the definitive practitioner's guide to this type of work.[6] Feathers defines legacy code simply as "code without tests." This definition is deliberately provocative. It shifts attention away from the age or programming language used for a system toward the question that actually matters: can you safely change this code? Code you can't verify is code you can't confidently modify, regardless of when it was written or what language it is written in.

Feathers' central approach is the "characterization test," an analysis that captures what code actually does rather than what it should do. Before changing anything, you write tests that document current behavior. These tests become a safety net. If your changes break something, the tests will catch it. This approach inverts the typical testing assumption. You are not testing against requirements; you are testing against behavior.

SpecOps applies a similar approach at a different level of abstraction. Where Feathers creates tests that capture behavior in code, SpecOps creates specifications that capture behavior in language domain experts can read and verify. A characterization test tells a developer "this function returns this specific output when given these specific inputs." A SpecOps specification tells a policy expert "this eligibility calculation applies the income threshold from Section 1902(a)(10)(A)(i)(VIII) of the Social Security Act, rounding income down to the nearest dollar." Both artifacts serve the same purpose: documenting what exists before attempting to change it.

Feathers also discusses the concept of "seams," places in the code where you can alter behavior without editing the code itself. Finding seams is essential for isolating components, getting them under test, and eventually replacing them. The SpecOps discovery phase does something analogous at the system level. Generating specifications naturally reveals component boundaries, dependencies, and integration points. These become the seams where teams can cut the system for incremental modernization. Chapter 8 will explore how this connects to the Strangler Fig pattern for replacing legacy systems.

The compilation mindset that SpecOps advocates is not a departure from established practice. It is an extension of what legacy code experts have known for twenty years: you can't **safely** transform what you do not understand. The contribution of SpecOps is recognizing that AI makes it practical to extract understanding at scale, and that specifications provide a form of understanding that domain experts, not just developers, can verify.

The Specification as the Valuable Artifact

Traditional approaches to AI-assisted modernization treat the generated modern code as the primary output of value. SpecOps takes a fundamentally different view: the specification is what matters most. More valuable than the legacy code. More valuable than the modern code that replaces it.[7]

This isn't just semantics. It changes what we optimize for, what we invest in, what we preserve. Consider why specifications capture institutional knowledge in ways that code can't. Legacy systems contain decades of accumulated business logic, policy interpretations, and edge case handling. This knowledge is embedded in code written by people who have long since retired, in languages few understand. When we generate a specification, we're not just documenting syntax. We're recovering and preserving institutional knowledge before it's lost forever.

Specifications enable human verification in ways code never can. Policy experts, program administrators, and business stakeholders can't easily verify that a Java translation of COBOL code is correct. But they can verify whether a specification accurately describes eligibility rules, benefit calculations, or tax logic. The specification creates a verification checkpoint where the people who actually understand what the system should do can confirm it's right. You're not asking them to review code. You're asking them to confirm that a description of the rules is accurate. That's something they can do better than anyone else.

As discussed in detail in Chapter 5, specifications provide optimal inputs to AI coding agents for code generation. A comprehensive specification serves as high-quality input for generating modern implementations. Rather than asking an AI agent to translate legacy code it may misunderstand, we're asking it to implement clearly articulated requirements. This is a task AI agents are demonstrably better at, and one where errors are easier to catch through testing against specified behavior.

Specifications are technology-agnostic in a way that code can never be. A well-written specification describes what a system does without being tied to how it's implemented. The same specification can guide multiple implementations: different technology stacks, different deployment models, different functional decompositions. The specification becomes a stable foundation that outlasts any particular technical implementation.

In twenty years, when the "modern" code needs modernization, the specification will still be valuable.

Specifications support incremental modernization over years-long efforts. Large-scale modernization typically spans years. A specification allows teams to modernize components incrementally while maintaining a single source of truth about system behavior. As new implementations replace legacy code, the specification ensures consistency and provides a reference for integration testing. This is how SpecOps integrates with patterns like the Strangler Fig pattern discussed in the next chapter.

And specifications create accountability in ways that scattered knowledge can't. When the specification is the authoritative source of truth, there's a clear artifact to review, approve, and maintain. Changes require updating the specification first, creating a review process that involves the right stakeholders. This is much harder to achieve when the "truth" is scattered across legacy code and in the heads of people who may or may not still work at the agency.

What does a good specification actually capture? When modernizing legacy government systems, specifications serve a different purpose than typical development documentation. They're not just implementation guides. They're artifacts that preserve institutional knowledge. This changes what "good" looks like.

A SpecOps specification must work for multiple audiences simultaneously: domain experts who verify that the specification captures policy intent, software developers and AI coding agents who need precision to generate correct implementations, and future humans who need to understand why the system behaves a certain way years from now, possibly after everyone currently involved has moved on.[8] That last audience is the one most specification formats neglect entirely.

What a Good Legacy System Specification Captures

Component	Description
Current system behavior	What the legacy code actually does
Current policy requirements	What the system should do per governing rules
Technical constraints	What the system can't do due to limitations
Known deviations	Documented reasons for policy/implementation gaps
Authoritative policy sources	Links to governing statutes and regulations
Decision records	Documentation from domain expert verification

Legacy system specifications can't just describe "what the system does." They need to distinguish between three different states. First, current system behavior: what the legacy code actually does today, bugs and all. Second, current policy requirements: what the system should do according to governing statutes and regulations. Third, technical constraints: what the system can't do regardless of policy, due to missing integrations or platform limitations. These three things can be in alignment or tension at any moment. And that alignment can shift over time without the code changing. A policy update tomorrow can transform compliant behavior into a violation.

Consider the example of a benefits system that should verify income against state tax agency records, but the legacy system only captures self-reported income because the integration with the tax agency was never built. A good specification would make this explicit: the policy requirement is income verification against tax records per the relevant directive; the current implementation uses self-reported income only; the deviation exists because no interface to the tax agency income verification service was ever built, with integration requested in 2019 but not funded; and the modernization note indicates the modern implementation should include tax agency income verification. This surfaces the gap, documents why it exists, and gives the modernization effort clear direction without pretending the legacy system does something it doesn't.

Government system specifications also need explicit links to authorizing statutes, regulations, or directives. Not just "these items are excluded from income calculations" but "per 42 USC § 1382a, the following items are excluded from income calculations." This is the "why" that survives personnel turnover. It's what allows future teams to evaluate whether behavior that was correct five years ago still aligns with current policy.

Testing the Concept

The SpecOps approach isn't theoretical. I built a demonstration using the IRS Direct File project, the free tax filing system that launched in 2024 and is now available on GitHub.[9] It's not a legacy system per se, but it's a good test case for several reasons. It has complex business logic interpreting the Internal Revenue Code. It has a multi-language codebase spanning TypeScript, Scala, and Java. And most importantly, it implements a set of rules that tax policy experts can verify.

To support the demonstration, I created reusable AI instruction sets for analyzing tax system code. These "skills files" included Tax Logic Comprehension as a foundation for understanding IRC references and tax calculations, Standard Deduction Calculation for extracting standard versus itemized deduction logic, Dependent Qualification Rules for capturing the tests for qualifying children and relatives, and Scala Fact Graph Analysis for understanding the declarative knowledge graph structures used in Direct File.

I pointed three different AI models at actual code samples from the Direct File GitHub repository and asked them to generate specifications. GPT-5 generated a specification for Standard Deduction that earned an A grade against my evaluation criteria. Gemini 2.5 Pro generated a Dependent Qualification specification that earned a B. Claude Sonnet 4.5 generated a Tax Bracket Calculation specification that earned an A-.[10]

All three models successfully extracted business logic into plain language suitable for domain expert review. A tax policy analyst could look at the generated specifications and say "yes, that's correct" or "no, you're missing the residency requirement." This is something they could not do, certainly not as easily, staring at raw software code.

Notably, these results came from single prompts without iteration. The skills I put together worked across different AI vendors, demonstrating the portability of the approach. There's clearly room for improvement through iteration and refinement, but even first-attempt results proved useful.

This matters for agencies facing modernization regardless of their timeline. SpecOps can be used whether immediate modernization is planned or still years away. There's never a bad time to aggregate and document knowledge about important government systems. Agencies can begin extracting specifications today, while institutional knowledge still exists and subject matter experts are still available. When modernization eventually happens, whether in two years or ten, agencies will have verified documentation, durable specifications, and a foundation for faster, less risky work.

The alternative is waiting until forced to modernize, then scrambling to reverse-engineer systems after the people who understood them have retired.

And the workforce clock is ticking. OPM data shows how acute this problem is: over 112,000 federal employees were added to the retirement rolls in FY 2025 alone.[11] This continues a pattern of roughly 100,000 federal retirements annually for over two decades. Each retirement potentially takes institutional knowledge about legacy systems out the door.

The IRS exemplifies the challenge. Facing both budget cuts and workforce reductions, over 30% of Chief Information Officer staff reportedly departed by mid-2025.[12] As one current IRS employee told reporters: "Systems break and it takes days to fix them because no one is left who knows how." The knowledge extraction work can't wait for a future modernization project. The experts may not be there when that project begins.

Redefining Success

When the specification becomes the source of truth, the definition of successful modernization changes. It's no longer just about producing working modern code.

Traditional modernization success means modern code that compiles, passes tests, and appears to work like the legacy system. SpecOps success means a comprehensive specification that accurately captures legacy behavior, verified by domain experts, with a modern implementation that correctly implements that specification and institutional knowledge preserved for future reference.[13]

This creates multiple success criteria, not just working code. Even if modernization is only partially complete, verified specifications represent real value delivered. Knowledge preservation has independent value from code translation. A specification without new code is valuable. New code without a verified specification is a gamble.

The role AI plays changes too. In transpilation, AI is the translator. Feed it legacy code, it produces modern code. The goal is to minimize human involvement in the translation process. In SpecOps, AI serves two distinct roles. First, understanding: analyzing legacy code to extract and articulate system behavior. Second, implementation: generating modern code from verified specifications. Human expertise bridges these roles through specification review and verification.[14]

When errors get identified and addressed changes as well. With direct translation, errors are discovered during testing if test cases happen to catch them, or in production when users encounter incorrect behavior. With SpecOps, errors get caught during specification review, when domain experts identify business logic mistakes before any code is written. The specification review checkpoint catches entire classes of errors: misunderstood business rules, incorrect policy interpretations, missing edge cases. Finding these errors in production can be catastrophic. Finding them during specification review is more manageable.

What happens when requirements change too? With direct translation, if requirements change, you modify the generated code directly. The code becomes the source of truth. Drift between what was intended and what exists becomes hard to track. With SpecOps, the specification is updated first. Changes are reviewed and approved by appropriate stakeholders. The updated specification then drives regeneration of affected code. There's a clear audit trail of what changed and why.

Breaking Free from the Legacy Trap

Even when direct translation succeeds in converting old code to new code, that new code becomes the foundation for a new legacy system. Technical debt starts to accumulate from day one. In twenty years, you'll face the same problem with Java code, or whatever was built, that you face with systems built with COBOL today.

Decoupling system behavior from technical implementation means governments never get stuck with decades-old legacy systems they desperately need to upgrade but don't fully understand.[15] The specification preserves institutional knowledge regardless of the underlying technology. When the next modernization is needed, the specification provides a foundation to build from. The knowledge doesn't have to be re-extracted from opaque code.

The specification outlasts implementations. Modern code is the current implementation. The specification is what endures. This is a different relationship between documentation and code than most organizations have experienced. It requires a mindset shift. But the payoff is systems that can evolve outside of periods of crisis.

The distinction between transpilation and compilation isn't just a technical nicety. It represents a fundamentally different answer to the question "what are we trying to accomplish?"

People working in government and civic tech have been doing this work the same way for decades. The methodology presented in the following chapters offers a fundamentally different approach.

The name "SpecOps" combines two proven approaches for building and managing modern software. "Spec-driven development" (discussed in Chapter 5) focuses first on comprehensive descriptions of system behavior before code is written. "GitOps" (discussed in Chapter 6) treats version-controlled representations of cloud infrastructure as the ultimate source of truth for an environment.[16] Like GitOps for system behavior, SpecOps means version-controlled specifications govern all implementations, creating an audit trail and enabling proper oversight of changes.

This reverses the typical government modernization pattern. Traditionally, when changes are needed, they are first made to the code with the hope of updating system documentation later. What happens in practice? Documentation falls behind, leading to the poor, incomplete, or non-existent documentation plaguing so many government legacy systems. With SpecOps, the specification is always the authoritative source. There's never a gap between system functionality and specification details.

The specification is what endures. The specification is what we can verify. The specification is what has value. Everything else is implementation details.[17]

The next chapter translates this conceptual foundation into a practical methodology, showing the six phases of SpecOps from Discovery through Deployment.

Chapter 8: The SpecOps Methodology

Two teams are modernizing similar benefits eligibility systems. Both have AI tools. Both have legacy COBOL code. Both face the same deadline.

Team A takes the direct translation approach. They point their AI tools at the COBOL and ask for Java. Within weeks, they have modern code. It compiles. It runs. They deploy it. Six months later, they discover the system has been miscalculating benefits for edge cases involving seasonal employment. The error was in the original COBOL, a bug that had been manually compensated for by caseworkers who knew to check those cases. The AI faithfully translated the bug into Java. Nobody caught it because nobody could verify whether the Java code correctly implemented required policies. They could only verify whether it matched the legacy COBOL.

Team B takes a different approach. Before writing any modern code, they use AI to generate a specification describing what the eligibility system does. A policy analyst who has worked with the program for twenty years reviews that specification. She catches the seasonal employment issue immediately: not because she can read COBOL or Java, but because she can read the specification's plain-language description of the eligibility rules and recognize that it doesn't match the policy. The specification gets corrected. Only then does Team B generate modern code, and the code implements the correct policy from the start.

The difference between these teams isn't the AI tools they used or the programming languages involved. The difference is methodology: the sequence of steps, the checkpoints for human verification, the recognition that understanding must precede implementation.

This chapter describes that methodology.

The Shape of the Process

SpecOps consists of six phases that form a coherent whole. Understanding the overall shape helps clarify how the phases connect and why this particular sequence matters.

The six phases are: Discovery and Assessment, where teams understand what they're working with; Specification Generation, where AI extracts and articulates system behavior; Specification Verification, where domain experts confirm the specification is correct; Modern Implementation, where code gets generated from verified specifications; Testing and Validation, where teams verify the implementation matches the specification; and Deployment and Knowledge Transfer, where the system moves to production and knowledge is preserved.[1]

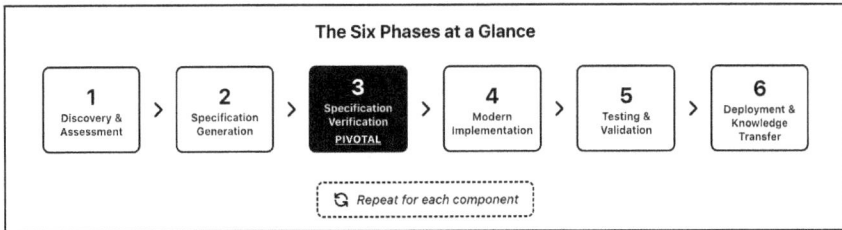

The Six Phases at a Glance

| 1 Discovery & Assessment | > | 2 Specification Generation | > | 3 Specification Verification PIVOTAL | > | 4 Modern Implementation | > | 5 Testing & Validation | > | 6 Deployment & Knowledge Transfer |

↻ Repeat for each component

The sequence isn't arbitrary. Each phase produces something the next phase requires. Discovery produces understanding of what exists and who knows about it. Specification Generation produces a draft description of system behavior. Verification produces confidence that the description is correct. Implementation produces code that realizes the verified description. Testing produces evidence that the code matches the specification. Deployment produces a working system and preserved knowledge.

Skip a phase, and you undermine everything downstream. Generate code without verification, and you may implement the wrong behavior confidently. Verify without proper discovery, and you may miss stakeholders who would have caught errors.

Phase 3, Verification, is the pivotal moment in the methodology. Everything before it builds toward producing something domain experts can evaluate. Everything after it flows from their validated understanding. This is where SpecOps diverges most sharply from direct translation approaches. Direct translation goes straight from legacy code to modern code, skipping the verification checkpoint entirely. SpecOps inserts a human-readable specification between them, creating space for domain expertise to catch errors that would otherwise propagate into the new system.

Though the phases are sequential in logic, they're iterative in practice. Teams don't specify an entire system, then verify it all, then implement it all. They work component by component: select a component, move it through all six phases, deploy it (often alongside the legacy system), then select the next component and repeat. This component-by-component approach enables incremental progress, manages risk, and delivers value continuously rather than only at the end of a multi-year effort.[2]

Discovery: Understanding What You Have

You can't modernize what you do not understand. Discovery establishes the foundation by mapping the system landscape, identifying who holds knowledge, and assessing what's at risk of being lost.

Discovery answers fundamental questions. What components make up this system? Who understands how they work? What documentation exists, and how reliable is it? Where is knowledge most at risk of being lost? What are the boundaries and dependencies between components? Without answers to these questions, teams make poor decisions about where to start, who to involve, and what to prioritize.

The most important output of Discovery isn't a technical inventory; it's a knowledge map. Who knows what? How available are they? When are they retiring? Which components have no one who understands them? This knowledge map drives prioritization. Components where expertise is about to walk out the door may need attention before technically simpler components. A system that's well-documented but stable may be lower priority than one that's poorly understood and actively degrading.[3]

Discovery also identifies natural boundaries in the system: places where components can be separated and modernized independently. Good boundaries have clear interfaces, limited dependencies, and well-defined data flows. Poor boundaries have tangled integrations that make independent modernization difficult. Finding these seams determines how work can be divided and sequenced.

AI can assist Discovery by analyzing code structure and generating preliminary documentation. But Discovery is fundamentally about organizational knowledge: understanding context, relationships, and risk that aren't encoded in source code. This phase requires human investigation: interviews, document review, relationship building.

Specification Generation: Making Knowledge Explicit

This is where AI becomes a force multiplier. AI agents analyze legacy code and produce draft specifications that describe system behavior in human-readable terms. The output isn't code; it's knowledge made explicit, ready for human verification.

In direct translation, AI reads legacy code and writes modern code. The transformation is opaque; there's no artifact that humans can evaluate to determine whether the AI understood correctly.[4] In SpecOps, AI reads legacy code and writes a specification. This specification is the intermediate artifact that makes verification possible. It describes what the system does in terms domain experts can understand: eligibility rules, calculation formulas, decision logic, edge cases.[5]

Specification vs. Documentation

These terms are often used interchangeably, but in SpecOps they mean different things. **Documentation** is descriptive. It describes what a system actually does, for better or worse. Documentation might note that "the system calculates income by multiplying the highest monthly earning by 12"—an accurate description of the code's behavior, even if that behavior is wrong.

Specifications are authoritative. They define what a system should do. A specification states that "income shall be calculated by averaging actual earnings across the employment period"—the correct policy, regardless of what the current code does.

Documentation follows code. When code changes, documentation gets updated to match. Specifications lead code. When policy changes, the specification gets updated first, then code changes to implement it.

In SpecOps, the specification is the contract. Code implements the specification. Tests verify conformance to the specification. When code and specification disagree, the specification is right and the code needs fixing.

This distinction matters because legacy systems often contain behavior that was never correct, or was correct once but policy has since changed. Documentation would faithfully describe that incorrect behavior. A verified specification captures what the system should do, giving teams a target to implement rather than a history to replicate.

AI is remarkably good at pattern recognition and explanation. Given legacy code and appropriate instruction sets, AI can identify business logic embedded in technical implementation, describe decision trees in plain language, document data transformations and calculations, flag edge cases and exception handling, and note uncertainties and ambiguities. What AI produces is a draft, a starting point for human review, not a finished artifact.

AI agents perform better with guidance. Instruction sets—sometimes called "skills" or "custom instructions"—teach AI about the specific context: COBOL idioms, domain terminology, specification format, what to emphasize and what to skip. Good instruction sets produce specifications that are structured consistently, use appropriate terminology, and focus on business logic rather than technical plumbing.[6]

Before domain experts see the specification, engineers review it for technical accuracy. Does the specification correctly describe the code's structure? Are the data flows accurate? Are dependencies identified? This technical review catches errors that domain experts wouldn't notice: misunderstandings about how the code works, not whether what it does is correct. Technical review and domain verification are complementary, not redundant.[7]

What AI can't do is verify that specifications are correct. AI can describe what code does, but it can't know whether what the code does matches policy intent. It can't catch bugs that have been in the system for decades. It can't recognize when behavior that seems correct is actually a workaround for a problem that no longer exists. This is why Verification exists as a separate phase.

Specification Verification: The Critical Checkpoint

Verification is the heart of what makes SpecOps different. Domain experts review specifications to confirm they accurately describe system behavior and policy intent. This checkpoint catches errors that would otherwise propagate into the new system undetected.

Direct translation approaches ask engineers to verify that modern code behaves like legacy code. But this only catches translation errors; it can't catch errors that existed in the legacy system or misunderstandings about what the system should do. SpecOps asks domain experts to verify that specifications correctly describe intended behavior. This catches a different class of errors entirely: business rules stated incorrectly, policy misinterpretations embedded in legacy code, edge cases that should be handled differently, bugs that have been compensated for manually, and behavior that was correct when written but policy has since changed.[8]

Domain experts can't read COBOL. They can't verify Java. But they can read a specification that says "Applicants with seasonal employment income have their annual income calculated by multiplying their highest monthly earnings by 12" and immediately recognize that this doesn't match the policy, which requires averaging actual earnings across the employment period. The specification creates a verification surface that domain expertise can engage with. This is the insight that makes SpecOps work.

Verification catches errors that no amount of testing against legacy behavior would reveal. AI hallucinations: business rules that don't exist. AI misunderstandings: rules stated incorrectly. Legacy bugs: behavior that was never correct. Outdated behavior: rules that should have changed but didn't. Omissions: important cases not captured. Every one of these would have propagated into the modern system if verification were skipped.[9]

Verification isn't just error detection; it's institutional knowledge capture. When domain experts review specifications, they provide context that wasn't in the code. "We round down here because of the 1997 policy memo." "This exception exists because of a court case in 2003." "Caseworkers know to double-check these cases manually." This context gets captured in the specification, transforming it from a description of code behavior into institutional memory.[10]

Skipping verification to save time is the most consequential mistake teams can make. Every error not caught in verification becomes an error in implementation, discovered later when it's far more expensive to fix, or worse, not discovered until it affects real people receiving incorrect benefits, paying incorrect taxes, or being wrongly denied services. The time invested in verification is repaid many times over in errors avoided downstream.

Implementation: Building from Verified Understanding

With a verified specification in hand, implementation becomes a more tractable problem. AI generates code from clear requirements rather than attempting to translate code it may not fully understand. The specification provides both the input for generation and the standard for evaluation.

Asking AI to translate COBOL to Java requires AI to understand what the COBOL does and express it in Java simultaneously. Errors in understanding become errors in translation, and distinguishing between them is difficult. Asking AI to implement a verified specification separates the concerns. Understanding has already happened in Specification Generation and been validated in Verification. Implementation is now purely about expressing that understood behavior in modern code. This is a task AI handles very well. Given clear requirements, AI generates reasonable implementations. Given ambiguous legacy code, AI generates plausible but potentially wrong translations.[11]

The specification defines what the implementation must do. This clarity enables engineers to evaluate whether generated code is correct, test cases to be derived directly from requirements, discrepancies between implementation and specification to be identified and resolved, and future changes to be evaluated against a clear baseline. Without a specification, "correct" means "matches legacy behavior." With a specification, "correct" means "implements verified requirements."

Traditional modernization asks: "Does the new system behave like the old system?" This question is hard to answer comprehensively and doesn't catch bugs that exist in both systems. SpecOps asks: "Does the new system implement the specification?" This question is easier to answer because specifications enumerate requirements, and it catches legacy bugs because the specification describes correct behavior, not legacy behavior. When new code differs from legacy behavior, the specification clarifies whether the difference is a bug to fix or a correction to celebrate.

Once you have a verified specification, you have options. You can use AI to generate implementation code, as this chapter describes. But you could also hand the specification to human developers and have them build the modern system traditionally. You would be walking away from the efficiency benefits of AI-assisted development outlined in Chapter 5, but the knowledge preservation value remains intact. The specification captures institutional knowledge regardless of how the code gets written.

This is worth emphasizing: the core value of SpecOps lies in the verified specification, not in using AI for implementation.

AI-assisted implementation is faster and often produces good results, but it's not essential to the methodology. What's essential is that you understand what you're building before you build it, and that understanding is verified by people who know the domain.

Even with verified specifications, implementation requires human oversight. AI-generated code needs review for quality, security, and maintainability. Test coverage needs to be comprehensive. Performance needs to be adequate. Implementation produces a candidate for deployment, not a finished system.

Testing, Deployment, and Knowledge Preservation

The final phases ensure the implementation is correct, transition it to production safely, and preserve the knowledge captured throughout the process. The specification remains valuable long after deployment.

Testing in SpecOps is specification-driven. Each requirement in the specification should have corresponding tests. Each edge case should be exercised. Coverage is measured against the specification, not against code paths. This changes what testing accomplishes. Instead of verifying that the code does something, testing verifies that the code does what the specification says it's supposed to do.[12]

Comparing new system behavior to legacy system behavior provides additional validation. Differences should be explainable: this difference corrects a legacy bug (the specification documents it), this difference reflects updated policy (the specification was verified against current policy), or this difference is unexpected (investigate). Legacy comparison catches errors that specification-based testing might miss and builds confidence that the modernization is comprehensive.

For critical systems, deployment rarely means instant cutover. Chapter 1 described the graveyard of failed big-bang modernizations: The Government of Canada's Phoenix Payroll disaster, the VA's repeated EHR failures, California DMV's $185 million across two abandoned attempts. A common thread in these failures was attempting to replace everything at once, flipping a switch from old to new and hoping for the best.

SpecOps, integrated with incremental modernization, takes a different approach. Modern and legacy systems coexist during a transition period. Some traffic routes to the modern system, some to legacy. Behavior is compared and validated. Traffic gradually shifts as confidence builds. Legacy is retired only when modern code is fully validated. This incremental approach manages risk and allows course correction.[13]

After deployment, the specification remains valuable. New team members learn the system by reading specifications, not reverse-engineering code. Changes begin with specification updates, reviewed by domain experts, then implemented. Auditors and regulators can understand system behavior without reading code. Future modernization efforts start with documentation, not archaeology. The specification outlasts any particular implementation. Code will be replaced again someday, but the specification persists.[14]

Integration with Incremental Modernization

SpecOps is designed to work with incremental legacy modernization approaches like the Strangler Fig pattern, enabling large systems to be modernized incrementally rather than all at once. The specification serves as both documentation of the legacy system and contract for its replacement.

Large systems can't be modernized in a single effort. The risk is too high; big-bang replacements fail spectacularly. The timeline is too long; business needs change during multi-year projects. The investment is too large; funding and political support rarely persist. Incremental modernization addresses these challenges by delivering value continuously, managing risk through small changes, and allowing course correction based on learning.[15]

The Strangler Fig pattern, named after vines that gradually envelop and replace their host trees, modernizes systems by building new components alongside the legacy system, routing some functionality to new components while legacy handles the rest, gradually expanding new component capability, and eventually retiring legacy when fully replaced. The challenge: how do you know what the new component needs to do? How do you verify it behaves correctly?[16]

SpecOps addresses this challenge by providing documentation of what each component does, a contract defining what the replacement must do, a test oracle for verifying equivalence, and an integration specification documenting how components interact. Without specifications, Strangler Fig relies on hope: hope that the team correctly understood legacy behavior. With specifications, Strangler Fig has verified understanding to build from.

Each component moves through all six phases. Discovery identifies the component's boundaries, dependencies, and knowledge holders. Specification Generation produces a description of component behavior. Verification confirms the description is correct. Implementation creates the modern replacement. Testing validates the replacement against the specification. Deployment transitions traffic from legacy to modern. Meanwhile, other components continue running on legacy. The system evolves incrementally, component by component, until the legacy system is fully replaced.[17]

Multi-year modernization efforts create challenges: domain experts retire, priorities shift, teams change. The specification provides continuity. Knowledge captured early remains available years later. New team members can understand components specified before they arrive. Changing priorities can be accommodated by reordering component modernization. The specification is institutional memory that doesn't retire.

A Component Through the Methodology

Abstract methodology becomes concrete when illustrated with a specific example. Consider an income verification module from a benefits eligibility system. It's selected for modernization because it has moderate complexity, clear boundaries, and an available domain expert, a policy analyst who understands the eligibility rules and is retiring in eight months.

During Discovery, the team identifies the income verification module as a 15,000-line COBOL program with interfaces to four external data sources. Documentation is sparse, but the policy analyst understands the business rules. The module has clear boundaries: it takes applicant information in and produces verified income figures out. A good seam for incremental modernization.

During Specification Generation, AI analyzes the COBOL and produces a forty-page specification describing income verification rules, data source integrations, calculation logic, and exception handling. The specification describes rules like: "If applicant reports self-employment income, request tax returns for the previous two years and calculate average monthly net income." Engineers review for technical accuracy and identify several ambiguities to clarify.

During Verification, the policy analyst reviews the specification. She catches two misstatements immediately: the specification says income is verified against the previous year's tax return, but policy requires two years for self-employment cases. She also identifies a seasonal adjustment calculation that the AI described incorrectly. More importantly, she provides context: "We round income down, not to the nearest dollar, because of a 1997 policy memo that's never been formally incorporated into the manual." This context is captured in the specification.

During Implementation, with the verified specification, the team generates a Python microservice. The code is clear because the requirements are clear. Engineers review for quality and security. Tests are derived directly from specification requirements.

During Testing, every requirement in the specification has corresponding tests. Legacy comparison reveals three differences: one is the corrected policy misstatement (the specification is right, legacy was wrong), one is the seasonal adjustment fix, and one is a performance improvement (same result, faster). All differences are explained and documented.

During Deployment, the new component is deployed alongside legacy. Ten percent of new applications route to the modern system initially. After two weeks with no issues, traffic increases to fifty percent, then one hundred percent. The legacy component runs in parallel for three months as a fallback, then is retired. The specification is finalized and archived.

The phases connect. Each produces something the next requires. Verification catches errors that would have propagated. The specification captures knowledge that would have been lost when the policy analyst retires. The component is modernized safely, incrementally, with verified understanding.

The Methodology's Promise

The two teams from this chapter's opening had the same tools and similar systems. The difference was methodology: whether they inserted a verification checkpoint between legacy analysis and modern implementation. Team A's direct translation produced code quickly but embedded errors that weren't discovered until they affected real people. Team B's specification-driven approach took longer upfront but caught errors before they became problems.

SpecOps is Team B's approach, systematized into six phases designed to capture knowledge, verify understanding, and implement with confidence. The specification is the artifact that enables verification—the compilation product that Chapter 7 described, the intermediate representation that makes domain expertise actionable.

Each phase serves a purpose. Skipping phases undermines what follows. The methodology works component by component, enabling incremental modernization of systems too large to replace at once. And at the center, the verification checkpoint ensures that domain experts, the people who actually understand what the system should do, have the opportunity to confirm that understanding before anyone writes a line of modern code.

The next chapter examines the principles and tools that make this methodology work: why specifications serve as source of truth, why domain experts are the arbiters of correctness, why changes must flow through specifications, and the tools that support the work. These principles guide adaptation when the standard phases don't quite fit.

Chapter 9: Core Principles, Practices, and Tools

A colleague once gave me the clearest definition of DevOps I've ever heard. "All DevOps really does," they said, "is make iteration cheap. That's it." No evangelising an elaborate framework or tools, no twelve-step process. It just makes iteration cheap.

That framing stuck with me because it explained why DevOps worked when other approaches often didn't. When deploying changes is expensive and risky, you batch them up and release quarterly (or even yearly!), which means you're always behind user needs and any given release is big, complex, and risky. When deploying changes is cheap, you release continuously, learn from real usage, and course-correct as you go.[1] The same work happens either way. But cheap iteration means you can refine your way to correctness instead of trying to get everything right the first time.

This principle shows up everywhere once you see it. Cheap iteration in manufacturing enables lean production. Cheap iteration in design enables rapid prototyping. Cheap iteration in writing enables multiple drafts. The expensive-iteration version of each of these processes looks completely different from the cheap-iteration version, even though the end goal is the same. Which brings us to creating software specifications.

Creating a comprehensive specification for a legacy system has traditionally been expensive. Prohibitively expensive in many cases. You need experts who understand the legacy code base, experts who understand the business domain, and enough time for all of them to work through every component, every edge case, every undocumented behavior. Even if you have the people (which is often a big if), the calendar time required can be enormous. And because iteration is expensive, there are strong incentives to try to get it right the first time, which you inevitably don't. Which, incidentally, is why most legacy systems have documentation that's either nonexistent, incomplete, or just plain wrong.

AI changes the existing cost structure for this work. An AI coding assistant can analyze legacy code and generate a draft specification in hours, not months. That initial draft will likely have some errors. It will miss nuances. It will need refinement. But refinement is now cheap. You can generate a draft, have a domain expert review it, identify what's wrong, regenerate with corrections, review again, and repeat until the specification is accurate. The same work happens either way—someone has to verify that the specification is correct—but cheap iteration means you can refine your way to correctness instead of trying to capture everything perfectly on the first pass.

This is what the SpecOps principles are designed to harness. Not AI as a magic oracle that produces perfect output, but AI as a tool that makes iteration cheap enough that refinement becomes practical. The principles that follow from this insight—specifications as source of truth, domain expert verification, changes flowing through specifications—are the practices that make cheap iteration productive rather than chaotic.

The Six Principles at a Glance

Principle	Why It's Important
The specification is the source of truth	Not the legacy code, not the new implementation. The specification is what persists.
Knowledge preservation precedes code translation	Capture what the system does before building its replacement.
Domain experts are the arbiters of correctness	The people who understand policy verify the specifications, not the code.
AI assists, humans verify	AI handles scale and pattern recognition. Humans provide judgment and domain knowledge.
Changes flow through specifications	All modifications start with specification updates, reviewed and approved before implementation.
Specifications should be accessible	If the verifiers can't understand it, verification doesn't work.

The lessons learned from failed modernization projects of the past, read in this light, tell the same story from the other direction. They describe expensive-iteration approaches: massive requirements gathering efforts that tried to capture everything upfront, documentation created once and never updated, verification deferred until it was too late to make changes. These past failures weren't for lack of effort or competence. They were structural consequences of iteration being too expensive to do properly.

SpecOps is, in some sense, the cheap-iteration version of legacy system modernization. The principles that guide it are the principles that make cheap iteration work.

The Specification as the Source of Truth

The most important principle in SpecOps is also its most distinctive: the specification, not the legacy code or the new implementation, is the authoritative description of what the system does and should do.

As discussed in Chapter 6, in infrastructure management GitOps established a powerful pattern: treat a Git repository as the single source of truth for infrastructure state. What's declared in the repository is what should exist in production. If there's a discrepancy, the repository wins. This declarative approach transformed how organizations manage complex cloud infrastructure.

SpecOps applies the same principle to system behavior. The specification repository is the authoritative source of truth about the system. The legacy code is data to be analyzed. The modern code is an implementation to be generated. But the specification is what persists, what gets verified, what guides both understanding and implementation.[2]

This principle has several important implications. First, it determines what gets preserved. When the specification is authoritative, institutional knowledge captured in a specification survives regardless of what happens to the code. Languages fall out of fashion, frameworks become unsupported, platforms get decommissioned. The specification is what endures.

Second, this principle enables verification by non-technical experts. Policy specialists and business analysts can verify that a specification correctly describes eligibility rules. They can't verify that Java code correctly implements those rules. The specification creates a checkpoint where the people who actually understand what the system should do can confirm that it's right.

Third, specifications support technology transitions. A well-written specification describes what a system does without being tied to how it's implemented. The same specification can guide implementations in different languages, frameworks, or architectures. The specification becomes the stable foundation that outlasts any particular technical choice.[3]

Knowledge Preservation Precedes Translation

Before generating a single line of modern code for a new component, invest in creating and verifying a comprehensive specification. The specification is not a means to an end; it's a valuable deliverable in its own right.[4]

Government agencies face a specific, urgent problem: the people who understand legacy systems are retiring, and their knowledge is walking out the door with them. An estimated 10,000 Baby Boomers retire every day.[5] Those that work in government often carry institutional knowledge that exists nowhere else. That's why the work to capture and compile this knowledge is so critical.

SpecOps takes this seriously by making knowledge capture the first priority, not an afterthought. The specification captures what the system does in a form that persists after the experts who understand it have moved on. Even if modernization stalls or changes direction, the specifications remain valuable.

Consider a scenario where you invest in generating and verifying specifications for a legacy system, but budget cuts prevent the actual modernization from proceeding. With traditional approaches, you have nothing to show for that investment. The requirements documents, if they exist at all, are outdated artifacts disconnected from the running system.

With SpecOps, the verified specifications are valuable on their own. They serve as training material for new staff. They provide documentation that auditors and oversight bodies can use. They clarify what the system actually does versus what people think it does. They inform maintenance and troubleshooting of the legacy system while it remains in operation. The knowledge has been preserved, regardless of what happens next.[6]

The idea of prioritizing understanding of a system is similar to themes in Marianne Bellotti's excellent book *Kill It with Fire*. In it, she argues that legacy systems deserve our respect precisely because they've survived for so long.[7] A system running for decades has been delivering value for decades. The system's longevity is evidence of success, not failure. The implication for modernization is significant: before replacing a system, you need to understand what made it succeed. Otherwise you risk building something that fails in ways the old system had already solved. SpecOps operationalizes this insight by making knowledge capture the first deliverable, not an afterthought.

The sequencing of steps matters. Knowledge preservation precedes code translation. This isn't just good practice; it changes the dynamics of a modernization project. When generating a specification is foundational, you invest in understanding before you invest in building. You verify correctness before you commit to implementation. You create checkpoints where domain experts can catch errors before they become embedded in code.

Direct translation approaches skip this step. There's typically no checkpoint for verification. No one can confirm that the AI understood correctly. No artifact persists that captures what was learned. You might get "working" code, but you haven't preserved knowledge.

Domain Experts as Arbiters of Correctness

The people who understand policy, business rules, and program intent are best positioned to verify the correctness of system behavior. SpecOps creates artifacts and processes that enable their meaningful participation in system modernization projects.

Traditional modernization approaches create a verification gap. The new code must correctly implement the behavior of the old code. But verifying this requires understanding both the legacy code and the new code. If your legacy system is written in COBOL and your new system in Java, you need someone who deeply understands both to verify correctness. Such people are rare and expensive. More often, verification depends on testing, which can only confirm behavior for the cases you think to test.[8]

SpecOps routes verification through specifications written in plain language. Policy experts, program administrators, and business stakeholders can't easily verify that Java code correctly implements COBOL logic. But they can verify whether a specification accurately describes eligibility rules, benefit calculations, or tax logic. They know what the system is supposed to do. They can read a specification and say: "Yes, that's right" or "No, that isn't how that is supposed to work."

This changes who can participate in verification. Instead of requiring rare technical experts who understand both legacy and modern code, verification involves the people who understand the business domain. These people exist in every government agency. They're the ones who answer questions when the system produces unexpected results. They're the ones who know why certain calculations work the way they do. SpecOps puts their knowledge to work.[9]

Domain expert verification isn't automatic. It requires specifications written at the right level of abstraction, detailed enough to be precise but not so technical that business experts can't engage with them. It requires a process that brings domain experts into the review cycle and gives them the time and mandate to actually review. It requires treating their input as authoritative, not advisory.

The domain expert doesn't need to know COBOL. They don't need to know Java. They need to know the business rules, and they need specifications that express those rules in terms they understand.

AI Assists, Humans Verify

AI excels at analysis and generation; humans excel at judgment and verification. SpecOps harnesses both.

In SpecOps, AI assists with two distinct tasks. First, AI analyzes legacy code to extract behavior patterns. It reads COBOL, or whatever legacy language you're working with, and generates explanations of what the code does. This leverages AI's ability to process large volumes of code and identify patterns. Second, AI generates new code from clear specifications. Given a verified specification, AI can produce a modern implementation. This leverages AI's ability to translate requirements into working code.[10]

Both tasks play to AI's strengths: processing at scale, pattern recognition, code generation. Neither task relies on AI's weaknesses: judgment about business correctness, understanding of organizational context, verification of accuracy.

AI will confidently generate specifications that are plausible but wrong. It will miss edge cases, misinterpret business logic, and produce output that looks correct but isn't. This is well-documented and unavoidable with current technology. It may be unavoidable with any technology.

Human verification is the essential checkpoint. Specifically, domain expert verification of specifications. This is where SpecOps differs most sharply from direct translation approaches. Direct translation has no verification checkpoint that domain experts can participate in. SpecOps is designed around exactly this checkpoint.

Neither AI nor humans alone is sufficient. AI without human verification produces fast but unreliable results. Human analysis without AI assistance is too slow and depends on rare expertise that's disappearing. The combination works because each party does what it does best. AI handles scale and pattern recognition. Humans provide judgment and domain knowledge. The specification is the artifact that bridges them.

Changes Flow Through Specifications

All system modifications begin with specification updates. The specification is not just documentation; it's a change management tool.

In a SpecOps environment, you don't modify code in isolation. You modify the specification first. The specification change is reviewed, verified by appropriate domain experts, approved, and then implemented. This ensures that changes are understood before they're built and that the right people have verified that the change is correct.[11]

This mirrors how infrastructure changes work in mature GitOps environments. You don't SSH into servers and make changes. You update the declared state in the repository, and automation propagates that change to a cloud environment. The same principle applies to system behavior: update the specification first, then update the implementation.

When changes flow through specifications, several things become possible. There's a clear audit trail of what changed, when, and why (and by whom). The specification repository shows the evolution of system behavior over time. Appropriate stakeholders review changes before implementation. Policy changes get reviewed by policy experts. Technical changes get reviewed by technical experts. And the system documentation stays current. The specification isn't an artifact that drifts out of sync with the code; it's the source of truth about system behavior that the code follows.

This means specifications are living documents, not static artifacts created once and filed away. They evolve as the system evolves. Every bug fix, every feature enhancement, every policy change flows through the specification first. This is a discipline that requires organizational commitment, but it pays dividends in maintained documentation, clear change history, and verified correctness.

Specifications Should Be Accessible

A specification should be readable by non-technical stakeholders while remaining detailed enough to guide implementation. This may require different views for different audiences.

Technical specifications often become impenetrable to non-technical readers. They accumulate jargon, implementation details, and structural complexity that serves technical needs but excludes domain experts. This creates a problem for SpecOps: if domain experts are the arbiters of correctness, they must be able to read and understand the specifications.[12]

SpecOps specifications can be structured in layers. A summary captures the essential behavior in plain language. Detailed requirements specify exactly what the system does. Technical annotations guide implementation. Different audiences engage with different layers. Domain experts verify the summary and detailed requirements. Engineers use the technical annotations. Everyone works from the same underlying truth, just at different levels of detail.

The default voice of a SpecOps specification is plain language. Technical terms are defined. Business logic is expressed in terms domain experts use. Implementation details appear in clearly marked sections that non-technical reviewers can skip. The goal is that someone who understands the business domain can read the specification and verify it, even if they've never written a line of code.

Overview of the SpecOps Toolchain

SpecOps requires specific types of tools, but doesn't demand exotic technology. Start simple, add sophistication as your practice of this method matures.

The toolchain falls into several categories: a specification repository for storing and versioning specifications, AI coding agents for analyzing legacy code and generating specifications, instruction sets that guide AI behavior for your specific context, and verification tools and processes for domain expert review.[13]

The minimum viable toolchain is straightforward: a Git repository for specifications, an AI coding agent, basic instruction sets, a markdown editor, and a pull request workflow for reviews. This is enough to get you started. You don't need sophisticated automation or specialized platforms. You need version control, AI assistance, and a review process.

The SpecOps Toolchain

Tool	Purpose
Git repository (GitHub, GitLab, Bitbucket)	Storing specifications with version control and collaboration
AI coding agent (Claude, GitHub Copilot, Cursor)	Analyzing legacy code and generating specification drafts
Basic instruction sets	For your legacy platform and domain, even rough ones you'll refine through use
Markdown editor (VS Code, Sublime, or any text editor)	Authoring specifications
Pull request workflow	Reviews and approvals, built into your Git platform

Don't wait for perfect tooling. Start with what you have. Improve as you learn.[14] The specification repository is where specifications live. Git provides version control, collaboration, and an audit trail. GitHub, GitLab, or similar platforms work well. Every change to a specification is tracked, attributed, and reversible. Directory structure should mirror system components. Naming conventions matter and should be established early.

AI coding agents are the technology at the heart of SpecOps. Claude, GitHub Copilot, Cursor, and similar tools analyze legacy code and generate specification drafts. They also assist with implementation once specifications are verified. Different tools have different strengths; the choice depends on your specific needs and constraints.

Instruction sets guide AI behavior. These are custom instructions that teach AI about your specific codebase, your legacy platform, your domain terminology. They're what transform generic AI capability into context-specific assistance. A well-crafted instruction set for understanding COBOL works whether you're modernizing a benefits system in California or a tax system in New York. The legacy language patterns are similar even if the business domains differ.[15]

Instruction sets develop iteratively. When you start, you likely won't have all the instruction sets you need. This is normal. You create basic instructions, use them, see where they fall short, refine them, and repeat. The instruction sets you end with will be substantially better than the ones you start with.

Verification and review tools support the domain expert review process. Pull request workflows, commenting systems, approval mechanisms. The tools should make it easy to assign reviews, track status, and maintain an audit trail.

As you mature, you can add automated validation to check specifications for completeness and consistency, testing infrastructure for verifying implementations, CI/CD pipelines for regenerating code when specifications change, and documentation sites for publishing specifications to broader audiences. These enhance the process but aren't required to start.

Chapter 13 will provide detailed guidance on setting up your toolchain, including specific tool recommendations, configuration guidance, and practical tips for getting started. This overview introduces the categories so you understand what's needed; that chapter will help you actually do it.

Building for Long-Term Sustainability

SpecOps is designed for the long term. The artifacts it creates, the processes it establishes, and the knowledge it preserves are meant to outlast any particular project or technology choice.

Today's modern code becomes tomorrow's legacy system. This is inevitable. Languages fall out of fashion. Frameworks become unsupported. The people who built the system move on. SpecOps creates specifications that persist regardless of implementation technology. A specification describing eligibility rules doesn't change because you move from Java to Ruby, or from on-premises to cloud, or from monolith to microservices. The business logic is the same. The specification captures it.[16]

Bellotti's analysis of modernization failures points to a deeper pattern. Organizations approach legacy modernization as a technical problem: old code needs to become new code. But the failures she documents in her book are attributable mainly to organizational dynamics, not technical limitations.[17] The processes, incentives, and knowledge structures that created the original system shape what replacement is possible. SpecOps addresses this by making knowledge preservation explicit and by routing verification through domain experts who understand the organizational context in which a system operates, not just the code.

When the next modernization comes, you don't start from scratch analyzing mysterious code. You start with verified specifications that describe what the system does. The knowledge has been preserved in a form that's technology-independent and human-readable.

Every specification verified is institutional knowledge captured. Every component modernized using SpecOps leaves behind documentation that supports future maintenance, future enhancement, future modernization. The investment in these specifications compounds. Organizations that practice SpecOps consistently will build a library of system knowledge that becomes more valuable over time.

The people who will modernize today's new systems haven't been hired yet. They may not even be born yet. SpecOps is, in part, an act of consideration for those future workers. You capture knowledge not just for today's project but for the projects that will follow in ten or twenty or thirty years. You write specifications that will still be readable and useful when the technologies we use today are as dated as COBOL is now.

The Principles as a Whole

The six principles of SpecOps form a coherent whole. Specifications are the source of truth because that's what enables knowledge preservation, domain expert verification, and long-term sustainability. Knowledge preservation precedes translation because capturing what the system does is more valuable than translating code. Domain experts are the arbiters of correctness because they're the ones who know what the system should do. AI assists while humans verify because that's the division of labor that plays to each party's strengths. Changes flow through specifications because that's how you maintain accuracy over time. And specifications should be accessible because verification only works if the verifiers can understand what they're reviewing.

These principles aren't arbitrary. They emerge from decades of lessons learned about what makes modernization succeed or fail. They're the reaction to the patterns of failure catalogued earlier in this book.

Bellotti's *Kill It with Fire* reaches similar conclusions from a different starting point. The experience she describes wrestling with legacy systems led her to emphasize understanding before acting, incremental replacement over big-bang rewrites, and respect for the institutional knowledge embedded in working systems.[18] SpecOps builds on these important insights by providing a methodology that makes them operational.

But there's a question we haven't fully addressed: why would these principles work especially well in government? Government has unique constraints, unique challenges, and unique requirements. The next section of this book will make the case that SpecOps isn't just compatible with government context but is particularly well-suited to it.

Part IV

Why SpecOps Works for Government

"Ignorance is the curse of God,
Knowledge the wing wherewith we
fly to heaven"

Henry VI; Part 2 - Act 4, Scene 7

Chapter 10: Knowledge Preservation and Domain Expertise

We met Tom in Chapter 1. He had worked at his agency for 32 years, understood the payroll system inside and out, and tried to document what he knew in his final two weeks. It didn't work. Six months after his retirement party, the agency called him back as a consultant to solve a problem no one else could diagnose.

Tom is now 72. The agency has been paying his consulting rate for three years. They've also tried, repeatedly, to extract what he knows into a form that would survive his eventual departure. Documentation projects. Knowledge transfer sessions. Video interviews. Mentoring assignments with younger staff.

None of it has worked. The documentation captures what Tom can articulate when asked, but that's not where his most valuable knowledge lives. Ask Tom to explain how the payroll system works and he'll give you an accurate but incomplete overview. Show Tom a specification that claims the system calculates overtime by multiplying hours over 40 by 1.5 times the base rate, and he'll immediately say "that's wrong for employees who started before 1987, they're grandfathered into the old calculation." The knowledge surfaces not through open-ended questions or interviews but through reaction to specific assertions of correctness. Tom can't inventory everything he knows, but he can recognize when something is incorrect. His expertise is most effective as a detector, not a generator.

The agency's problem isn't that Tom won't share. He wants to. The problem is that traditional knowledge transfer assumes knowledge is explicit, articulable, and transferable through conversation. Much of what makes Tom valuable is tacit, contextual, and only reveals itself in the presence of specific details about how the system is supposed to work.

This is the knowledge preservation challenge at its core. And it's a challenge that government faces more acutely than almost any other sector.

The Demographic Cliff

The federal workforce is older than the American workforce overall, and the gap is widening. The average federal worker is 47.2 years old, compared to a median age of 42.2 for the U.S. labor force.[1] Over 28 percent of full-time permanent federal workers are age 55 or above, compared to about 23 percent in the private sector.[2] Only about 7 percent of federal employees are under 30, compared to nearly 20 percent in the general labor force.[3]

The age distribution has shifted dramatically over the past two decades. In 2005, 8.1 percent of federal workers were 60 or older while 6.9 percent were in their 20s. By 2024, nearly twice as many workers were in their 60s (14.8 percent) as in their 20s (7.8 percent).[4] The federal workforce isn't just older than the private sector; it's getting older faster.

This isn't exclusively a federal problem. Federal workforce data is pretty comprehensive, but state governments face similar demographic challenges. State employees who maintain unemployment insurance, Medicaid eligibility, and tax systems are aging alongside their federal counterparts. The COBOL programmers that New Jersey desperately sought during the pandemic weren't federal employees; they were state workers, many of whom had retired years earlier. At least 45 states still run COBOL systems, and the workforce maintaining those systems reflects similar age distributions.[5] The analysis in this chapter uses federal data, but the same patterns apply across governments at all levels.

The departures of federal employees in 2025 accelerated these trends dramatically. Approximately 317,000 federal employees left government in 2025, compared to typical annual separations of around 150,000 in other years.[6] After accounting for approximately 68,000 new hires, the net reduction in the federal workforce was roughly 249,000, representing roughly 11 percent of the federal workforce departing in a single year.[7] The departures included a mix of retirements, voluntary resignations, early retirement incentives, and involuntary separations. Many agencies saw 30 percent or more of specific technical teams depart.

Even before the 2025 disruptions, roughly 100,000 federal employees retired annually. OPM data shows over 112,000 federal employees added to retirement rolls in fiscal year 2025 alone, continuing a pattern stretching back to fiscal year 2000.[8] The predicted "retirement wave" that observers have warned about for decades may have finally arrived, compounded by policy-driven reductions in the workforce.

For legacy systems specifically, the workforce situation is dire. The Government Accountability Office found that for an at risk DOD contract management system, the average age of developers and subject matter experts "is above 60, putting the system at significant risk in being able to support and maintain it into the future." Officials stated that it is "difficult to find COBOL and assembly language code developers and the learning curve once they are identified is also significant."[9]

The expertise to maintain these systems is not just aging but actively departing. Each retirement potentially takes institutional knowledge about legacy systems out the door.

Why Government Is Different

The knowledge preservation problem is more acute in government than in the private sector for several important reasons.

Federal employees tend to stay longer than private sector workers. A 30-year tenure is not uncommon in government but rare in private industry. Longer tenure means more accumulated knowledge concentrated in individual people. It also means fewer opportunities for knowledge to spread across multiple people through turnover. The same factors that make government employment stable create knowledge concentration risk.

Government systems encode highly specialized domain knowledge that typically doesn't exist outside government. Tax law, benefits eligibility, regulatory compliance, procurement rules. When the person who understands TANF eligibility rules retires, you can sometimes find contractors who know the domain, but they're drawing from the same shrinking pool of expertise. Tom's agency brought him back as a consultant, and firms like COBOL Cowboys maintain networks of retired government programmers for exactly this purpose. But these experts are in their 60s and 70s. The consultant market for legacy government systems isn't a renewable resource; it's a retirement community. You can rent expertise for a while, but you can't build institutional capacity by depending on people who are at the end of their careers. The knowledge either transfers to permanent staff or it eventually disappears when even the contractors are gone.

Understanding legacy systems requires two kinds of knowledge: technical skills like COBOL and mainframe architecture, and domain knowledge like policy rules, program requirements, and business logic. These don't need to reside in the same person, but they need to connect somehow. Traditional approaches struggle because the people with technical skills can't evaluate whether business logic is correct, and the people with domain knowledge can't read the code. But both kinds of expertise are aging out of government, and both are hard to replace. The challenge isn't just finding technical skills or domain expertise; it's preserving the accumulated understanding that connects them.

Structural barriers make knowledge transfer particularly difficult in government. Hiring processes are slow; backfilling departed expertise can take months. Budget constraints limit overlap periods where departing and arriving staff can work together. Classification and pay systems make it hard to bring back retirees or hire specialized consultants. Security clearances for sensitive systems can't be transferred or accelerated. By the time a replacement is hired, cleared, and trained, the departing expert is often long gone.

The Gap Between Policy & Implementation

Government systems encode decades of accumulated policy decisions, legislative changes, and regulatory interpretations. This complexity makes government business rules particularly difficult to understand, document, and preserve.

Government programs change through legislation, regulation, and administrative interpretation. Each change adds to the complexity of existing systems. A benefits calculation that may have started simple in 1970 can be modified by dozens of policy changes over many years. The existing code often reflects years of accumulated decisions, each reasonable in its moment. Understanding the current system requires understanding its entire history.

Between statute and code lies interpretation. What counts as "income"? What's a "household"? When does eligibility begin?

These questions have answers embedded in the code, often undocumented. The answers were provided by people who understood both the policy and the system at the time. That combination of understanding is exactly what retires with departing experts.

The complexity manifests differently across domains. Tax systems contain IRC references, deduction logic, and filing status rules built up over decades. Benefits systems encode eligibility determinations with cascading rules and exceptions. Regulatory systems reflect years of rulemaking and enforcement decisions. Each system contains its own universe of domain knowledge embedded in code that nobody fully documents because nobody could possibly document all of it.

Specifications as Institutional Memory

SpecOps specifications capture knowledge in a form that persists beyond individual employees, that can be verified by domain experts, and that serves multiple purposes over time. They are institutional memory in durable form.

Specifications capture what other artifacts don't. Code shows what a system does; specifications explain why. Documentation describes intended behavior; specifications capture actual behavior. Process manuals describe how to use systems; specifications describe how systems work. Specifications bridge the gap between technical implementation and business meaning.

Unlike tacit knowledge that depends on a person being present to share it, specifications are durable. A verified specification survives the retirement of the person who verified it. The knowledge captured in the specification doesn't depend on anyone's memory. When questions arise years later, the specification provides answers that can be consulted, searched, and referenced.

Specifications enable verification loops that transfer knowledge by requiring explicit articulation. Current domain experts verify specifications against their understanding. When those experts retire, their verification is recorded in the specification itself. New domain experts can review existing specifications and update them as policies change, and before systems are updated. The verification process creates a chain of accountability that preserves institutional understanding across generations of staff.

The same specification serves multiple purposes over time. Training: new staff learn systems by reading specifications. Troubleshooting: when systems behave unexpectedly, specifications clarify correct behavior. Auditing: oversight bodies can review specifications to understand system logic. Modernization: future projects start with verified documentation rather than archeology. Compliance: specifications document that systems implement policy correctly.

Traditional documentation is created once and then usually decays. Specifications in SpecOps are maintained as the source of truth. Changes to systems require specification updates first. The specification stays synchronized with reality because the process requires it. This is the fundamental difference between documentation as afterthought and specification as foundation.

The Case for Acting Now

The window for capturing institutional knowledge is closing. Rapidly. The experts who can verify specifications are leaving. Beginning knowledge capture now, even without immediate modernization plans, creates value that can't be created later. There is never a bad time to do this work.

AI can generate draft specifications from code, but only humans can verify that those specifications correctly describe what the system does and should do. Those humans are the domain experts who understand the business rules, the policy requirements, and the historical context. They're the people who are retiring. Every month that passes, more potential specification verifiers leave. Unlike code translation, which can theoretically happen anytime, verification requires specific people. Those people are a diminishing resource.

Traditional thinking ties documentation to active modernization projects. This is a mistake. Specification work has independent value regardless of modernization timeline. Agencies can and should pursue knowledge capture even when no modernization project is currently funded, when funding is being sought but hasn't arrived, when modernization is planned but years away, or when there's uncertainty about when or whether modernization will happen. Waiting for a modernization project to begin knowledge capture means starting that project without the experts who could have informed it.

There is never a bad time to begin capturing institutional knowledge. Every verified specification represents value delivered, regardless of what happens next. If modernization proceeds, the specifications accelerate it. If modernization stalls, the specifications still support maintenance, training, and operations. If the experts retire before modernization begins, the specifications preserve what they knew. The work compounds: each specification verified makes the next one easier, as patterns emerge and context accumulates. Starting "too early" has no downside; starting "too late" could mean permanent knowledge loss.

Even if modernization is years away, verified specifications have immediate value. They support maintenance of existing systems by documenting how things actually work. They enable training of new staff who can read specifications even if they can't read legacy code. They clarify what systems actually do versus what people assume they do. They create audit trails for compliance and oversight. They reduce key-person risk by distributing knowledge beyond the experts who currently hold it. They provide a foundation for estimating modernization costs and timelines when funding discussions do occur. The work is valuable whether or not it leads immediately to new code.

Agencies that wait will face modernization pressure eventually. System failures, security incidents, compliance requirements, changing policies, or simply the inability to make needed changes will force action. When that pressure arrives, the experts who could have verified specifications will be gone. Reverse-engineering systems without expert verification is slower, riskier, and less reliable. More importantly, some knowledge simply can't be recreated once the people who hold it are gone.

The IRS exemplifies both the challenge and the consequences of delay. The agency faces legacy system challenges dating back decades, and significant workforce departures have compounded the difficulty. Over 30 percent of Chief Information Officer staff reportedly departed by mid-2025.[10] Current employees describe the impact directly: "Systems break and it takes days to fix them because no one is left who knows how."[11] The knowledge loss is already affecting operations, not just future modernization. This pattern will repeat across government as departures continue. Agencies watching the IRS situation should ask themselves: are we capturing knowledge now, while we still can?

Enabling Non-Technical Verification

One of the most important features of SpecOps for government is that it enables verification by people who understand policy but don't write code.

Traditional modernization creates a verification gap. Verifying that new code correctly implements old code requires understanding both. The people who can read both COBOL and Assembly are rare. Testing can verify behavior for the cases you think to test, but can't confirm that you've thought of all the important cases. Policy experts who know what systems should do can't verify code directly. Technical experts who can read code don't know what systems should do. The verification gap means errors can persist undetected until they affect real people.

Specifications bridge this gap by expressing system behavior in terms that policy experts can evaluate. The statement "the system calculates income by summing wages, tips, and interest" is verifiable by non-programmers. A policy analyst can confirm: "Yes, that's correct" or "No, self-employment income should also be included." The verification happens in natural language, not in code. Domain expertise, not programming expertise, becomes the qualifying skill.

This dramatically expands the pool of potential verifiers. Every agency has policy experts who understand program rules. Most agencies don't have many people who can read legacy code. SpecOps makes the larger pool of employees useful for verification. More potential verifiers means more thorough verification. It also means verification can continue as technical experts retire, because verification doesn't depend on technical skills.

Specifications create explicit checkpoints in the modernization process. At each checkpoint, the right experts can confirm correctness. Errors are caught before they're embedded in code. The cost of fixing errors in specifications is far lower than fixing errors in deployed systems. The checkpoints also create accountability: someone has signed off that this specification is correct. If problems emerge later, there's a record of who verified what and when.

The practical implications reshape how modernization teams work. Teams can include policy analysts, not just programmers. Verification sessions become conversations about business rules, not code reviews. Domain experts feel ownership of the process because they can meaningfully participate. The modernization effort becomes a collaboration across expertise domains rather than a technical project that domain experts observe from a distance.

Tom's consulting engagement won't last forever. He's 72, and even the most dedicated consultant eventually stops taking calls. The agency knows this. They've tried everything they can think of to capture what he knows before he's gone for good.

What they haven't tried is what SpecOps makes possible: having Tom verify specifications rather than generate documentation. Instead of asking Tom to explain everything he knows, show him specifications generated from the code and ask whether they're correct. Let his pattern-matching expertise flag what's wrong. Let his accumulated knowledge surface through agreement (or disagreement) with explicit assertions. The question shifts from "tell us what you know" to "is this right?"

The specifications that Tom verifies will persist after he stops consulting. They'll capture his expertise in a form that future staff can use, that auditors can review, that modernization teams can build on. The knowledge won't depend on Tom being available for a phone call.

This is what knowledge preservation looks like in practice. Not knowledge transfer sessions that try to extract expertise through conversation. Not documentation projects that capture what people can articulate when asked. But a process that creates explicit, verifiable artifacts that encode institutional understanding in durable form.

The federal workforce is older than it's ever been. The departures of 2025 accelerated a trend that was already concerning. The people who understand how government systems work are leaving, and the traditional approaches to preserving their knowledge aren't working.

SpecOps offers something different: a way to capture knowledge before it walks out the door, to verify that capture with the experts while they're still available, and to preserve the results in a form that serves the institution long after individuals have moved on.

The window for this work is closing. The time to begin is right now.

Chapter 11: The Politics of Modernization

In 1999, I was working as a tax policy analyst for the State of Delaware. My job was supposed to involve writing legislative analyses, calculating revenue projections, and advising other parts of our department on policy implementation activities. Instead, for several chaotic months, I found myself doing something entirely different: creating spreadsheets to double-check whether the state's new payroll system was calculating paychecks correctly.

Delaware had embarked on an ambitious legacy modernization project called PHRST (Payroll Human Resources Statewide Technology), an effort to replace a collection of fragmented payroll systems with a modern, integrated PeopleSoft implementation-one that would work for all employees. The system would handle payroll for all state agencies, school districts, and charter schools. It was exactly the kind of modernization that government technology leaders dream about: consolidating disparate systems, creating a single source of truth, and enabling self-service capabilities for employees.

Unfortunately, the system was launched prematurely. Everyone involved knew it wasn't ready. But no one wanted to deliver the bad news to the Governor's office or the Budget Office that the rollout needed to be delayed.[1] Political timelines had been set. Commitments had been made. So the system went live, and everyone crossed their fingers and hoped for the best.

What followed was chaos. Calculation errors emerged. Some employees received incorrect paychecks. The people who should have been doing their actual jobs were pulled into belated testing. We built spreadsheets to verify the calculations the payroll system was supposed to be handling automatically. Business analysts across the state became a human verification layer for a system that was supposed to reduce manual work.

Three years later, then Governor Ruth Ann Minner acknowledged what those of us on the inside had experienced first-hand. "After years of delays and cost overruns, we successfully implemented PHRST," she said in her 2002 State of the State address.[2] The word "successfully" was doing a lot of work in that sentence. The system eventually stabilized, but only after more resources were poured in, more oversight was established, and years of workarounds were institutionalized. Delaware is still running that system today, more than twenty-five years later, and is only now undertaking another modernization effort to replace it.[3]

That experience shaped how I think about government technology modernization. The technical problems were real, but they weren't the fundamental issue. The fundamental issue was political: no one could say "not ready" to the people who needed to hear it. The incentive structure made it easier to launch and fail than to delay and explain.

Why Government Modernization Is Different

Government IT modernization operates under constraints that don't exist in the private sector. These constraints aren't incidental or the result of poor management. They arise from the nature of democratic governance itself.

In the private sector, IT failures are business problems. They may be expensive, embarrassing, and career-limiting for those responsible. But they remain internal matters, resolved through internal processes. In government, IT failures become political scandals. Healthcare.gov's launch problems in 2013 nearly derailed the Affordable Care Act itself. Canada's Phoenix pay system became a national embarrassment that eroded public trust in government competence. The political cost of visible failure vastly exceeds the political benefit of invisible success.

This asymmetry creates powerful incentives for risk aversion. But paradoxically, it also creates pressure to launch systems before they're ready, because political timelines don't align with technical requirements.

Administrations change every four to eight years. Major modernizations can take longer than that. Budget cycles are annual; modernization roadmaps span multiple years. Political announcements create commitments that technical teams must meet regardless of actual readiness. In Delaware, the rollout date was fixed by political commitment, not technical assessment. When those two things conflicted, political commitment won.

Government also requires accountability mechanisms that add complexity in ways the private sector doesn't face. Procurement rules designed to ensure fairness create rigidity. Audit requirements demand documentation that slows progress. Oversight committees require briefings that consume leadership attention. Each layer of accountability is individually reasonable. Collectively, they create an environment where the very mechanisms meant to prevent failure often contribute to it.

An Information Problem

Political leaders must make decisions about technology projects they can't directly evaluate. This information asymmetry creates the conditions for the kinds of failures we see repeatedly with legacy modernization projects.

Governors, legislators, and agency heads approve projects they can't technically assess. They rely on briefings, summaries, and assurances from people who have their own incentives to present optimistic pictures. The people closest to technical reality are furthest from political power. The people with political power have the least visibility into technical reality.

Delivering bad news up the chain is career-limiting. In Delaware, no one wanted to tell the Governor's office that the system wasn't ready. Each layer in the hierarchy softens the message. By the time information reaches decision-makers, it's been filtered through multiple rounds of optimism. "We have some challenges but we're addressing them" becomes the universal status update, regardless of actual project health.

Demonstrations to leadership compound the problem. They are designed to show what works, not what doesn't. Proof-of-concept systems look impressive but can hide integration problems. Political leaders see polished presentations, not the fragility behind them. The gap between demonstration and production deployment is where projects die.

When political timelines slip, testing is almost always what gets compressed. You can't delay the announcement. You can only delay the parts no one sees. Testing is invisible to leadership; launch dates are visible. Delaware's "belated testing" with spreadsheets was a symptom of this compression. The testing that should have happened before launch happened after, performed by people like me who should have been doing other work.

The Parallel with Phoenix

The Delaware PHRST experience is far from unique. As discussed in Chapter 3, the same dynamics produced similar outcomes in Canada's Phoenix pay system, but at a much larger scale.

The Government of Canada launched Phoenix, a new pay system for federal employees, in 2016. Like Delaware's PHRST, it was built on PeopleSoft. Like PHRST, it was launched despite warnings from technical staff that it wasn't ready. Political commitments to timeline and cost savings overrode technical concerns. The result: hundreds of thousands of federal employees were paid incorrectly. Some went months without paychecks. Others were overpaid and faced demands for repayment they couldn't afford. Eight years later, the system still hasn't been fully fixed. Total cost: over $3.5 billion and counting.[4]

The Phoenix Pay System: A Case Study in Legacy Failure

Year	Event
2009	Cabinet approves Transformation of Pay Initiative with $310 million budget
2011	IBM awarded pay modernization contract
2009-2015	Software deferments, privacy breach, and rollout delays largely brushed aside
Early 2016	Unions petition government to delay launch until system is thoroughly tested
Feb 2016	Phoenix launched to 34 departments despite warnings; 1,200 pay advisors reduced to 550
April 2016	Thousands of employees report pay errors; 30% have incorrect paycheques
2017	Auditor General calls Phoenix an "incomprehensible failure of project management"
2018	Senate Committee calls system an "international embarrassment"; projects costs could reach $2.2 billion
2024	Government announces plan to replace Phoenix with new system
2025	Nine years later, 300,000+ pay transactions remain unprocessed; total costs exceed $5.1 billion

Phoenix and PHRST share structural similarities that go beyond both using PeopleSoft. Both had political timelines that couldn't move. Both had technical teams that couldn't deliver bad news effectively. Both launched prematurely rather than delay. Both required years of additional investment to stabilize. The pattern isn't bad luck. It's predictable given the existing incentive structure.

The long tail of these failures extends for decades. In Delaware, a pension calculation error from the PHRST era went undetected for 28 years. In 1997, the Compensation Commission had recommended changing how legislative pensions were calculated. Due to a clerical oversight during the system transition, the recommendation was never properly codified into law.[5] Although the pension office adopted the new calculation method, the legal code still reflected the old rules. When the error was finally discovered in early 2024, the state had to revert to pre-1997 rules. The result: $900,000 in retroactive payments and a $13,000 increase in monthly pension payroll.[6]

A clerical error from the chaotic PHRST implementation period, compounding silently for nearly three decades, ultimately costing nearly a million dollars to correct. That's the long tail of technical debt from troubled implementations. The system that "successfully" launched in 1999 created problems still being felt in 2025.

Accountability and the Need for Audit Trails

Government modernization must serve accountability requirements that go beyond mere functionality. Systems must be auditable, explainable, and traceable in ways that private sector systems often don't require.

Government systems handle public money and require public trust. Every decision must be defensible to auditors, legislators, and courts. "The system made that decision" is not an acceptable explanation when someone's benefits are denied or their taxes are miscalculated. Business logic must be traceable to authorizing policy. Errors must be explainable in terms non-technical stakeholders can understand.

Code fails this test. Code embeds business logic in implementation details that auditors can't read. When systems produce unexpected results, explaining why requires technical expertise that oversight bodies don't have. The gap between policy intent and system behavior is invisible when the only record of that behavior is source code.

Specifications serve accountability in ways that code can't. They document intended behavior in terms that non-technical stakeholders can verify. They create an audit trail from policy to implementation. Changes flow through specifications, creating traceable history. Auditors can compare specifications to authorizing policy without reading code. The specification becomes evidence of intent, not just implementation.

Version-controlled specifications show what changed, when, and why. Decision records document the rationale for choices. Domain expert sign-off creates accountability for verification. The specification repository becomes an institutional record that supports the oversight that government projects require.

Risk Aversion and How to Manage It

Risk aversion in government IT is rational given the track record of failures. The solution is not to demand more risk tolerance, but to structure work so that risks are smaller and failures are recoverable.

Given the failure rates documented throughout this book, caution is reasonable. Careers have ended over failed modernizations. The personal downside of failure exceeds the personal upside of success. Risk aversion isn't a lack of bravery or commitment; it's rational self-preservation in an environment where the incentives punish visible failure more than they reward invisible success.

Large projects concentrate risk into single points of failure. Years of investment hang on a single launch date. When that launch fails, everything else fails as well. This is how Healthcare.gov and Phoenix happened. The scale of investment made failure catastrophic, which made the pressure to launch despite problems even more intense.

Incremental delivery changes this dynamic. Small releases have small failure modes. The Strangler Fig pattern, discussed in Chapter 8, enables incremental replacement rather than big-bang cutover. Each increment can be evaluated before proceeding. Failures are local, not systemic. Success builds confidence for the next increment. Political leaders can see progress continuously rather than waiting years for a launch that may or may not work.

Making the Case to Leadership

Technical leaders must translate SpecOps benefits into terms that resonate with political decision-makers. This requires speaking to their concerns, not yours.

Political leaders care about risk, budget, timeline, accountability, and communication. Will this work? What will it cost? When will it be done? What happens if it fails? How do I explain this to stakeholders? These are reasonable questions that deserve direct answers:

- "This approach reduces risk by delivering incrementally. Instead of waiting years to see if something works, we can demonstrate progress in months. Each phase produces real value, even if later phases don't proceed. And because we verify specifications with domain experts before we build, we catch misunderstandings early—when they're cheap to fix."

- "The specifications we create capture institutional knowledge that outlasts any single implementation. We're not just building a system; we're documenting how your programs actually work. Even if modernization pauses, that knowledge remains valuable. And because we verify understanding before coding, we spend less time fixing things we built wrong."
- "You'll see working results in the first 90 days—not just planning documents, but actual functionality. Progress is continuous and visible, not a years-long wait followed by a launch that may or may not work. We can align our technical milestones with your political calendar."
- "Specifications create an audit trail that connects policy decisions to system behavior. Domain experts sign off on requirements before development begins, so there's clear ownership of what we're building. When someone asks why the system does what it does, we can point to documented decisions, not just code."

Specifications reduce verification risk specifically. Verifying specifications before coding catches errors early, when they are cheapest to fix. Domain experts can validate understanding without reading code. Errors discovered in specifications cost a fraction of errors discovered in production. The verification step that we improvised in Delaware with our spreadsheets becomes systematic rather than emergency response.

What would have been different in Delaware if specifications had existed? Leaders could have verified understanding before committing to a launch date. Technical concerns would have been visible in documented gaps rather than filtered through layers of optimism. The "ready or not" decision would have had evidence behind it.

Working Within Political Reality

Twenty-five years after those chaotic months in Delaware, I still think about what we were really doing with those spreadsheets. We were-in essence- creating specifications, in a crude and haphazard way. We were documenting what the system was supposed to calculate, then checking whether it actually did. We were building the verification layer that should have existed before the system launched.

The political dynamics that made Delaware's PHRST launch prematurely haven't changed. Elected officials still face political deadlines. It's still true that no one wants to deliver bad news up the chain of command. Budget cycles still don't match modernization timelines. These are structural features of democratic governance, not problems to be fixed.

What has changed is what's possible. AI-assisted specification development means we can build those verification layers systematically, not as an emergency response to a failed launch. We can create artifacts that political leaders can actually evaluate, not just code they have to take on faith. We can structure work so that the question "is this ready?" has an evidence-based answer.

The politics of modernization won't disappear. But we can work within those politics more effectively when specifications make progress visible, when incremental delivery reduces catastrophic risk, and when domain experts can verify understanding before a system launches.

My Delaware colleagues and I were doing primordial specification-driven development way back in 1999. We just didn't know it, and we were doing it way too late. The methodology this book describes is, in part, an attempt to institutionalize what we figured out under pressure and to do it before that pressure arrives.

Chapter 12: Collaboration as Force Multiplier

The logic seems obvious. Fifty states all need unemployment insurance systems. They all process claims, verify eligibility, calculate benefits, and issue payments. Why would each state build its own system from scratch? Multiply this across every function of government: licensing, permitting, benefits administration, case management, tax processing. The duplication is staggering. The federal government alone spends an estimated $12 billion annually on software, and lawmakers have long suspected that agencies are paying contractors to build systems that already exist elsewhere.[1]

This logic has driven two decades of policy. In 2016, the Office of Management and Budget issued M-16-21, the Federal Source Code Policy, requiring agencies to make custom-developed code available for sharing and reuse.[2] The policy established a pilot program requiring agencies to release at least 20% of new custom-developed code as open source. In December 2024, the SHARE IT Act strengthened these requirements, mandating that agencies share code and publish metadata so other agencies can find it.[3] Similar initiatives exist at the state level. The intent is clear: stop the duplication, share what works, save money.

And yet. The expected revolution in government code sharing hasn't really materialized. Agencies publish code to repositories that other agencies rarely use. States continue building their own systems despite the availability of alternatives. The open source repositories grow, but the duplication continues.

This isn't because government workers are stubborn or wasteful. It's because the premise, that government software is like Lego blocks that can be snapped together across jurisdictions, misunderstands what most government software actually is. Understanding why direct code sharing has disappointed expectations is the first step toward identifying what kinds of collaboration actually work.

Why Code Reuse Has (Largely) Disappointed

When someone suggests that California should adopt Texas's unemployment insurance system, they're imagining that the core functionality is the same and only superficial details differ. In reality, the differences run very deep. Each state has made distinct policy choices about eligibility rules, benefit calculations, waiting periods, work search requirements, and dozens of other parameters. These aren't configuration options sitting in a settings file. They're embedded throughout the codebase, in the validation logic, the calculation routines, the workflow rules, the data structures.

A licensing system for professional engineers encodes specific requirements about education, examination, experience, and continuing education that vary by state. A benefits eligibility system implements rules that reflect each jurisdiction's policy decisions about who qualifies for assistance under what circumstances. The "same" function across jurisdictions shares only surface-level similarity. The policy logic that makes each system unique is precisely what makes the software hard to share.

In principle, you could separate the jurisdiction-specific policy logic from the common infrastructure: the intake forms, document management, payment processing, notification systems. In practice, this separation is difficult architectural work that most systems weren't designed to support. Legacy systems especially tend to have policy logic intertwined with system infrastructure in ways that make extraction expensive and error-prone (which is why the word "monolith" is so frequently

167

used in describing these systems). Extracting the reusable parts requires understanding the system deeply enough to identify what's common versus what's jurisdiction-specific. This is exactly the kind of knowledge work that's in short supply.

Ironically, modern code may be harder to share than legacy code. COBOL is COBOL everywhere, which is part of why instruction sets for legacy platforms can be widely shared. But modern technology stacks are fragmented. One state may choose Ruby on Rails, another may use Python and Django, a third may go with Java and Spring, a fourth may standardize on the Windows stack. Even if two states have built functionally similar licensing systems, the implementation differences mean that adopting one requires expertise in that specific stack, including framework versions, deployment patterns, and dependency ecosystems. A state without Ruby developers can't readily adopt a Ruby application, regardless of how well-designed or well-documented it is. The technology choice that made sense for the originating agency can act as a barrier for potential adopters.

The economics of software development can work against reuse as well. There's an informal maxim in software development that designing systems for reusability makes them more expensive to build. Creating software that other organizations can adopt requires anticipating variation, designing clean interfaces, separating configuration from logic, documenting thoroughly, and maintaining backward compatibility as the system evolves. These are best practices that improve software quality generally, but they add cost and complexity. Government software development operates under constant budget pressure. Contracts reward delivering specific functionality on time and within budget, not building elegant abstractions that hypothetical future users might appreciate. There's no procurement mechanism that rewards an agency for spending an extra 30% to make their system reusable by another government. The benefits of reusability accrue to other agencies, possibly agencies that don't even exist yet, while the costs are borne by the team trying to meet their own deadlines. The incentive structure actively discourages building for reuse, even when everyone agrees reuse would be valuable in the abstract.

Even when code is available and potentially useful, adopting it isn't free. An agency considering another agency's code must evaluate whether it fits their needs, assess the quality and security, plan the adaptation work, and commit to ongoing maintenance. For an agency with limited technical staff, this evaluation burden alone can exceed the cost of building something simpler from scratch. The SHARE IT Act addresses discoverability by requiring metadata publication, but discoverability was never the primary barrier. The barrier is that evaluating and adapting code requires exactly the technical capacity that many agencies simply do not have.

Sharing That Works

The collaboration models that have succeeded in government share a common pattern: they provide value at layers of the stack where the problem is genuinely common, without requiring agencies to adopt jurisdiction-specific policy logic.

Login.gov provides authentication services that federal agencies can integrate without running their own identity infrastructure.[4] An agency doesn't need to understand how identity proofing works under the hood. They call an API, and Login.gov handles the complexity of security, compliance, and continuous improvement. Cloud.gov provides a FedRAMP-compliant Platform-as-a-Service offering.[5] Agencies deploy applications without managing the underlying infrastructure or navigating the compliance process independently. The service handles what is common so agencies can focus on what's specific to their mission. These services work because authentication and cloud infrastructure management are genuinely common problems. The policy logic that makes each agency's applications different sits above these services, not within them.

The U.S. Web Design System (USWDS) takes a different approach but follows the same principle.[6] USWDS provides design components, patterns, and guidance that agencies adopt to build accessible, consistent web experiences. It's not a running service but a set of standards and code that agencies incorporate into their own applications. USWDS works because the problems it solves, accessible typography, consistent navigation, mobile responsiveness, are common across agencies. The system doesn't try to dictate what content agencies publish or what workflows they implement. It handles the presentation layer and leaves the policy layer to each agency.

The pattern of what has worked seems clear. Successful government collaboration tends to happen at infrastructure and standards layers, below the policy-specific application logic. The further down the stack you go, the more common the problem becomes. Authentication is more common than benefits eligibility. Hosting is more common than case management. Design patterns are more common than business rules.

This pattern suggests where future collaboration should focus: identifying additional layers where problems are genuinely shared and investing in common solutions at those layers.

Collaboration Layers

Layer	Type	Examples	Sharing Model
Infrastructure	Shared services	Login.gov, cloud.gov	Consume via API
Standards	Common patterns	USWDS, API standards	Adopt and customize
Knowledge	Instruction sets	Legacy platform skills, domain patterns	Use and contribute
Application	Business logic	Benefits systems, licensing	Limited sharing expected

The Unexpected Opportunity

The Federal Source Code Policy, the SHARE IT Act, and similar state initiatives encouraging the publishing of open source software have succeeded at one thing: getting code published. Federal agencies and state governments have released code for benefits systems, licensing platforms, case management tools, tax systems, and dozens of other functions. Most of this code will never be picked up and run by another agency. But it still has tremendous value.

An instruction set that helps AI assistants understand COBOL becomes more robust when it's informed not just by textbook COBOL but by actual COBOL running in state unemployment systems. An instruction set for extracting business rules from legacy benefits systems gets better when it's been exposed to how multiple jurisdictions implemented similar concepts in different ways. The published code becomes raw material for building better instruction sets, discussed in detail in Chapter 5, even when it can't serve as directly reusable software.

This inverts the usual criticism of government open source efforts. If the traditional view is that governments have made all this code available and nobody's really using it, SpecOps provides a new way to look at it. Now, because of these efforts, governments have access to lots and lots of code and they can learn from at scale. The value isn't in directly running the code but in understanding it well enough to teach AI agents how to work with similar systems.

This suggests a collaboration model different from traditional code sharing. Rather than expecting agencies to adopt each other's applications, agencies that have successfully modernized legacy systems can contribute instruction sets informed by their experience. The instruction sets capture knowledge about technical patterns, legacy platforms, and domain concepts without embedding jurisdiction-specific policy.

A state that has worked through modernizing its unemployment insurance system can share instruction sets that help AI assistants understand the patterns common to unemployment systems generally: claim processing workflows, eligibility determination structures, benefit calculation approaches. Another state using those instruction sets still implements its own policy rules, but benefits from accumulated knowledge about the technical patterns. The laws and policies requiring agencies to publish custom-developed code have created the foundation for this kind of collaboration, even if that wasn't their original intent.

What Makes Instruction Sets Shareable

Instruction sets avoid the problems that limit traditional code sharing because they encode knowledge about platforms and patterns rather than jurisdiction-specific policy.

COBOL is COBOL whether you're modernizing benefits in California or taxes in New York. Mainframe patterns, database structures, batch processing approaches, and legacy language conventions don't vary all that much based on jurisdiction. An instruction set that teaches an AI assistant to comprehend JCL or analyze CICS transactions works across agencies because the technical substrate is common.[7] This is fundamentally different from application code, where the policy logic varies. Instruction sets operate at the technical layer where government systems are similar, not the policy layer where they diverge.

Instruction sets contain guidance and examples, not live data or proprietary business rules. They can be published with permissive licenses and shared freely without security or legal concerns that might limit code sharing. An agency can share an instruction set that helps AI assistants understand how benefits eligibility rules are typically structured without revealing the specific rules for their programs. The pattern knowledge transfers; the policy specifics can stay internal.[8]

Adopting another agency's instruction set doesn't require running their code or adapting their architecture. It means giving your AI assistant additional context about patterns it might encounter. If parts of a shared instruction set don't apply to your systems, you simply don't use them. The adoption burden is minimal compared to evaluating and adapting application code.

As more agencies contribute instruction sets, the shared library becomes more valuable. An agency starting a COBOL modernization project benefits from instruction sets contributed by every agency that has done similar work. The early contributors to this library create value for everyone who follows. This creates positive incentives for contribution. Unlike code sharing, where agencies may hesitate to publish work that others might criticize, instruction sets represent collective learning that reflects well on contributors.[9]

What Can Be Shared Freely

Instruction sets in the following categories should be shareable with minimal restriction:

- Legacy platform comprehension (COBOL, RPG, mainframe environments)
- Database and data structure patterns (DB2, IMS, VSAM)
- Specification structure standards (this is what a useful specification looks like)
- Modern implementation patterns
- Testing and validation approaches
- General domain patterns (benefits concepts, licensing workflows)
- Government compliance frameworks (Section 508, FISMA, FedRAMP)

What should typically stay internal: specific business rules for your programs, sensitive system architecture details, security implementations, and vendor-specific integrations.

A Collaboration Model That Fits Government

The collaboration opportunity isn't one-size-fits-all. Different layers of the stack call for different collaboration models. At the infrastructure layer, shared services like Login.gov and cloud.gov provide capabilities that agencies consume. At the standards layer, resources like USWDS provide patterns and components that agencies adopt. At the knowledge layer, shared instruction sets provide context that makes AI-assisted modernization more effective. At the application layer, code remains largely jurisdiction-specific, and that's appropriate.

This layered model acknowledges that government agencies have genuine differences in policy and mission while creating collaboration opportunities where commonality exists.

The SpecOps community model envisions a shared repository of instruction sets organized by category: legacy languages, legacy platforms, government domains, specification patterns, implementation approaches. Agencies contribute what they've learned; agencies starting new projects benefit from accumulated knowledge. The contribution model is lightweight. Publishing an instruction set that helped your team doesn't require ongoing support obligations. Users adapt shared instruction sets to their context and contribute improvements back if they choose.[10]

An agency beginning a legacy modernization effort no longer starts from zero. They can draw on shared instruction sets for COBOL comprehension, mainframe environment patterns, benefits domain concepts, or whatever applies to their situation. The AI assistants they're working with come pre-loaded with knowledge accumulated from across government. This doesn't make modernization easy. The hard work of understanding specific systems, verifying specifications with domain experts, and implementing modern replacements remains. But the starting point is higher. The learnings from one agency's modernization make the next agency's work a little easier.

The Force Multiplier

The disappointment with government code sharing stems from a mismatch between expectations and reality. We expected agencies to adopt each other's applications directly. But most government software embeds policy choices that make direct adoption impractical. The business logic problem, technology stack fragmentation, the economics of building for reuse, and the adoption burden all work against the simple vision of shared applications.

The collaboration opportunities that have worked operate differently. Shared services provide common capabilities at the infrastructure layer. Shared standards provide consistent patterns at the design layer. And now, shared instruction sets can provide accumulated knowledge at the layer where AI assistants learn about legacy systems.

None of this eliminates the need for each agency to do its own modernization work. California can't simply download Texas's unemployment system and run it. But California can benefit from instruction sets informed by Texas's experience modernizing COBOL, just as Texas can benefit from instruction sets informed by California's work with benefit systems.

The force multiplier isn't in sharing finished applications. It's in sharing the knowledge that makes building applications faster and more reliable. In a field where every agency has historically solved the same problems in isolation, that's a big deal.

Having established why SpecOps is a good fit for government projects, including the knowledge preservation imperative, the politics of modernization, and the collaboration opportunity, we turn next to practice. How do you actually get started? How do you work effectively with AI tools and the people on your team? What challenges should you expect, and how do you address them?

Part V

SpecOps in Practice

*"Things won are done; joy's soul
lies in the doing"*

Troilus and Cressida; Act 1, Scene 2

Chapter 13: Getting Started

The meeting had been on the calendar for a month, but now that everyone was actually in the room, the weight of the moment was palpable. The state's Deputy CIO, had just gotten approval for a pilot modernization project. She had six months to show results on a system that hadn't been significantly updated since the Clinton administration. The last person who deeply understood the COBOL code had retired two years ago. The Deputy CIO had read about specification-driven approaches and believed they could work for the challenges facing her team. Now she was staring at a whiteboard with that team, asking the question that would determine whether this attempt succeeded or failed: "Where do we start?"

Everyone in the room knew what was at stake. They'd all heard stories, and some had lived through their own. One tale of past failure that loomed large-California's DMV had canceled its modernization program in 2013 after seven years and $135 million spent. That was the second attempt at modernization, coming after a previous $50 million failure in 1994 that also lasted seven years.[1] Two attempts, fourteen years total, $178 million, and nothing to show for it. That shadow was hanging over the team now as they prepared their own attempt at legacy system modernization.

The Deputy CIO didn't want to be the next cautionary tale. But she also couldn't keep maintaining a system held together with luck and institutional memory that was rapidly disappearing. She had to modernize—the challenge was doing it without repeating the pattern of failure that has defined similar efforts for decades.

She looked at the whiteboard. Someone had written three questions on it:

- What system component do we start with?
- Who needs to be in the room?
- What do we need to have ready before we begin?

These are the right questions to ask before starting a legacy system modernization project. And this chapter will try to answer them.

Choosing Your Starting Point

The pilot component you select for legacy system modernization can shape everything that follows. Choose well, and you'll build organizational confidence, demonstrate value, and create momentum for the broader modernization effort. Choose poorly, and you may doom the initiative before it has a chance to prove itself.

This is not the time for the hardest problem. Nor is it the time for a trivial exercise that doesn't actually prove out an approach. You need something in the middle: complex enough to be meaningful, simple enough to complete, with the right characteristics to let SpecOps demonstrate its value to broader stakeholders.

The ideal pilot component has several characteristics. First, it should be moderate in complexity, something your team can complete specification and verification for in two to four weeks.[2] Complex enough that success means something, but bounded enough that failure doesn't sink the whole initiative. Second, it should have known behavior that domain experts can actually verify. If nobody understands what the component does, you can't validate whether your specifications captured it correctly. Third, and critically, domain experts must actually be available to do that verification. A component whose only expert retired last year is a poor choice, no matter how attractive it might seem otherwise.

Beyond these essentials, look for clear boundaries. Components with well-defined integration points make better pilots than those deeply entangled with the rest of the system. When testing whether your modern implementation works, you don't want to be debugging interface problems at the same time. Similarly, look for business value sufficient that stakeholders will care about the outcome. A successful pilot that nobody notices won't build the organizational momentum you need.

The prioritization process starts with understanding what you have. Conduct discovery across the legacy system to understand its components, their relative complexity, and the state of documentation and expertise for each. Assess risk factors: Where is knowledge most endangered? Which areas have the most technical debt? Then rank modernization priority based on which replacements would deliver the most value the soonest.

When you've identified candidates, apply the filter: moderate complexity, known behavior, available domain experts, clear boundaries, sufficient business value. The component that passes all these tests is your pilot.

Pilot Selection Checklist

- ☐ Moderate complexity (2-4 weeks to complete)
- ☐ Known behavior (verifiable by domain experts)
- ☐ Domain expert availability confirmed
- ☐ Clear boundaries with other components
- ☐ Business value sufficient to matter
- ☐ Success criteria defined
- ☐ Source code accessible (can you actually get to it?)
- ☐ Related materials available (JCL, copybooks, documentation)

Before committing, define what success looks like. What does "this pilot worked" mean in concrete terms? Establishing success criteria upfront prevents goalpost-moving later and gives everyone a shared target to aim for.

There are also some characteristics to avoid. The "too easy" trap catches teams who pick something trivial to guarantee success. A component so simple that anyone could modernize it proves nothing about SpecOps. You'll get your win but learn nothing and convince no one. The "crown jewels" mistake goes the other direction: staking everything on your most critical, most complex, most politically charged system. If something goes wrong, and something almost always goes wrong the first time, you've created a disaster rather than a learning opportunity.

Watch out for the "nobody cares" problem. Components that don't matter don't get resources, attention, or executive air cover when things get difficult. And watch out for the "expert desert," components where no one remains who can verify whether specifications are correct. Without verification, you're just generating documentation that might be wrong. Finally, avoid integration nightmares. Components so entangled with other systems that you can't test them independently will make the pilot far harder than it needs to be.

Building Your Team

SpecOps requires a different team composition than traditional development or even traditional legacy modernization efforts. The emphasis on specification generation, domain expert verification, and AI-assisted analysis creates unique staffing needs. Getting this mix right matters as much as selecting the right pilot component.

For a pilot project, the minimum viable team is eight to ten people.[3] That sounds like a lot, but the roles span distinct functions that can't easily be combined. You need a Program Manager working full-time to coordinate activities, remove blockers, and manage stakeholders. You need a Technical Lead, also full-time, who may also serve as your AI Engineering Lead initially. You need two to three Software Engineers with modern stack skills and the adaptability to work with AI tools. You need at least one QA Engineer focused on testing and validation.

On the domain side, you need a Lead Domain Expert with deep knowledge of the business area being modernized. This person will be the primary verifier of specifications, the arbiter of whether the AI correctly understood the system's behavior. You'll likely also need one or two Domain Specialists to provide additional coverage. And you need a Business Analyst who can bridge the technical and domain worlds, someone who can take complex information and make it clear and accessible.[4]

The Business Analyst role in SpecOps carries significant research responsibilities beyond traditional requirements gathering. This includes conducting user observations and interviews to discover undocumented features and workarounds, analyzing error logs and ticket histories to understand actual system behavior, studying historical records and policy documents to trace why systems evolved as they did, and investigating whether observed behaviors are bugs or intentional features. These research activities are essential for creating accurate specifications when documentation is incomplete or missing.

If your initiative extends beyond a pilot, the recommended team size is twelve to eighteen people. At that scale, you separate the Technical Lead and AI Engineering Lead roles, add more engineers for parallel work on multiple components, increase QA capacity, bring in more domain specialists for broader coverage, and add a dedicated DevOps engineer.[5] Larger teams may also benefit from having multiple Business Analysts, with one focused primarily on research activities: discovering undocumented system behavior through user observation, conducting historical investigation of policy evolution, and cross-referencing ticket histories and training materials to understand why systems behave as they do. This research-focused BA role becomes increasingly valuable as modernization scales to components with sparser documentation or where institutional knowledge has already been lost.

The most important shift in team composition compared to traditional legacy modernization is the change in what expertise matters most. Traditional approaches require staff who know the legacy language. If you're modernizing COBOL, you need COBOL programmers. These experts are rare, expensive, and many are retired or retiring.

SpecOps changes this equation. AI handles much of the legacy code analysis, reducing though not eliminating the need for legacy language expertise. What becomes more critical is proficiency with AI tools and, most importantly, domain expertise. The people who understand the business rules, the policy rationale, the edge cases that matter become more valuable than ever. The shift is from "need rare COBOL experts" to "need domain experts, and AI-skilled engineers."[6]

If you do have legacy language experts available, use them. Have them validate AI-generated specifications. Have them help design instruction sets that teach the AI about your specific codebase's patterns and idioms. Capture their knowledge while you still can. But if you don't have them, SpecOps can still work in ways that traditional modernization approaches can't.

Among all the roles, two deserve special attention. The AI Engineering Lead designs and refines instruction sets for AI agents, understands AI capabilities and limitations, troubleshoots when AI-generated content is poor, and builds team fluency with AI tools. This role can be combined with Technical Lead initially but becomes important to separate as the initiative scales.

The Lead Domain Expert may be the single most important role for SpecOps success. This person verifies specifications for correctness, provides authoritative answers on business rules, teaches domain knowledge to the rest of the team, and has the credibility with stakeholders to sign off that specifications accurately capture system behavior. Without an effective Lead Domain Expert, the entire verification process breaks down.

The Minimum Viable Team
8-10 people for a SpecOps pilot project

Leadership & Coordination

Program Manager (Gov't staff recommended)	Full-time

Technical Delivery

Technical Lead / AI Engineering Lead (Can be combined initially)	Full-time
Software Engineers (2-3 people)	Full-time
QA Engineer	Full-time

Domain Expertise & Verification

Lead Domain Expert (Gov't staff recommended)	75%+
Domain Specialists (1-2 people)	50-75%
Business Analyst	Full-time

For government agencies, the question of staff versus contractors inevitably arises. A hybrid approach works best. Government staff should fill roles requiring continuity and institutional knowledge: Program Manager, Product Owner, Lead Domain Expert, and some core engineering positions. These are roles where knowledge needs to stay with the organization after any contract ends. Contractors can provide specialized skills like AI expertise, additional engineering capacity during peak periods, and specific technical architecture knowledge.[7]

The critical discipline is ensuring knowledge transfer happens throughout the project, not as an afterthought at the end. Specifications stay with the government. Instruction sets get shared openly. Everything gets documented. Government staff pair with contractors so capability builds over time. The risk of a contractor-only approach is that knowledge walks out the door when the contract ends. The risk of a government-only approach is that you may lack specialized AI skills initially. Balance these risks through deliberate hybridization.

When assembling your team, watch for red flags. Technical staff who dismiss domain expert input will undermine the verification process that makes SpecOps work. Domain experts who won't engage with technical artifacts can't provide the verification you need. Leadership that wants "silver bullet" solutions will be disappointed when the work turns out to require actual work. Team members resistant to using AI tools will slow everyone down. Anyone who can't work collaboratively will create friction in a process that depends on close coordination between technical and domain perspectives. And perfectionism that prevents progress will kill momentum. Ship incrementally, improve continuously.[8]

Setting Up Your Toolchain

The tools you need for SpecOps are simpler than you might expect. You don't need exotic technology or expensive platforms. You need a handful of capabilities that most technology organizations already have access to, configured to support specification-driven work.

But before you think about tools, you need access to what they'll work on. This sounds obvious and is often harder than expected.

Source code access is the foundation. Where does the legacy code actually live? For mainframe systems, it might be in libraries accessible only through specific terminals or emulators. If you're lucky, it might be in a version control system; or if you're not it might be in backup tapes that haven't been accessed in years. Who controls access? What permissions are required? In large organizations, finding the authoritative source code and getting permission to work with it can take weeks. Better to discover this early than to have your pilot stall while waiting for approvals.

Beyond source code, you'll need related materials: JCL (Job Control Language) and job schedules for batch processing systems, database schemas and copybooks and data dictionaries, whatever system documentation exists however outdated, test data and test scripts if available, and interface specifications with other systems. Each of these may live in different places, controlled by different groups, requiring different approvals.

Anticipate access challenges. Security restrictions on production code are common and reasonable, and navigating them takes time. Approval processes move at their own pace. Code stored in proprietary formats may require special tools to extract. Documentation is often scattered across departments or lost entirely. And organizational politics around who "owns" the system can complicate even straightforward requests. Start the access process early, in parallel with other setup activities.

This matters for SpecOps specifically because AI agents need actual source code to analyze. If your access is incomplete, your specifications will be incomplete. Gaps in materials become gaps in understanding. Better to discover what you're missing before the pilot officially begins.

With access addressed, the minimum viable toolchain is straightforward: a Git repository for specifications, an AI coding agent, basic instruction sets for your legacy platform, a markdown editor for specifications, and a pull request workflow for reviews.[9] That's it to start. Don't get hung up on the notion of "perfect" tooling.

The SpecOps Toolset

Tool Category	Purpose	Examples	When Needed
Version Control	Store specifications, track changes, enable collaboration	GitHub, GitLab, Bitbucket	Day 1
AI Coding Agent	Analyze legacy code, generate specification drafts, assist implementation	Claude, GitHub Copilot, Cursor, Aider	Day 1
Instruction Sets	Guide AI behavior for your specific legacy platform and domain	Custom markdown files; community libraries	Day 1 (basic); refine ongoing
Markdown Editor	Author and edit specifications	VS Code, Sublime, any text editor	Day 1
Pull Request Workflow	Review and approve specification changes	Built into GitHub/GitLab	Day 1

Tool Category	Purpose	Examples	When Needed
Automated Validation	Check specifications for completeness and consistency	Custom scripts, linters	As team matures
Testing Infrastructure	Verify implementations match specifications	pytest, JUnit, Jest, BDD tools	Month 4+
CI/CD Pipelines	Regenerate code when specifications change, run validation	GitHub Actions, GitLab CI, Jenkins	As team matures
Documentation Site	Publish specifications to broader audiences	GitHub Pages, GitBook, MkDocs	As team matures

Each piece serves a specific purpose. Git provides version control, collaboration, and an audit trail. Every change to a specification is tracked, attributed, and reversible. The pull request workflow creates natural checkpoints where verification happens. AI coding agents, whether Claude, GitHub Copilot, Cursor, or others, are the core technology enabling SpecOps. They analyze legacy code, generate specification drafts, and assist with implementation. Instruction sets guide their behavior, teaching them about your specific codebase. Markdown keeps specifications readable by both humans and machines, without proprietary format lock-in.

Setting up the specification repository deserves some thought. Create a dedicated repository, separate from code repositories. Establish a directory structure that mirrors your system's components. Define naming conventions and stick to them. Set up branch protection so specifications can't be modified without review. Create templates for specifications so the format stays consistent.

For AI coding agent selection, consider your legacy languages, the context window size you need for your codebase, cost, and integration with your existing development environment. Multiple tools may be appropriate for different phases. GitHub Copilot now supports multiple underlying models including Claude.[10] Don't over-optimize this decision. Start with what's accessible and learn what works for your specific situation.

As your practice matures, add capabilities: more sophisticated instruction sets refined through experience, automated validation tools that check specifications for completeness and consistency, testing infrastructure for specification conformance, CI/CD pipelines that regenerate code when specifications change, documentation sites for publishing specifications to broader audiences. But these come later. Start simple.

One principle should guide tooling decisions: SpecOps tools should integrate with, not replace, existing development infrastructure. Use your existing source control, CI/CD systems, and issue tracking. Extend existing pipelines rather than creating parallel ones. The goal is to make SpecOps feel like a natural extension of existing practices, not a completely separate parallel system that competes for attention.

Creating Your First Instruction Sets

Instruction sets are how you teach AI agents to understand your specific legacy code and generate useful specifications. They're sometimes referred to as "skills" or "custom instructions."

Whatever you call them, they're among the most valuable and reusable components of the SpecOps methodology.

Of note, instruction sets are more portable than code. A well-crafted instruction set for understanding COBOL works whether you're modernizing a benefits system in California, a tax system in New York, or a licensing system in Texas. The legacy language patterns are probably similar even if the business domains differ. This creates enormous opportunity for collaboration and sharing across organizations.[11]

When you start a SpecOps project, you likely won't have all the instruction sets you need. This is normal and expected. Instruction set development happens in parallel with specification work, each informing the other. You'll create basic instructions, use them, see where they fall short, refine them, and repeat. The instruction sets you end with will be substantially better than the ones you start with.

Think of instruction sets at different maturity levels.[12] Level 1, basic instructions, are sufficient to start. They provide general guidance for the task, common patterns to look for, examples of desired output, and known pitfalls to avoid. Level 2, refined instructions, develop through use. They include specific patterns for common scenarios, detailed examples from actual experience, handling of edge cases, and integration with other instruction sets. Level 3, mature instructions, are battle-tested and comprehensive, with extensive pattern libraries, numerous examples and test cases, community validation, and documented version history.

Start at Level 1 and evolve toward Level 3 through practice. Don't try to create perfect instruction sets before beginning. Create good-enough instruction sets, use them, improve them, then repeat.

You'll need instruction sets in several categories. Legacy platform comprehension instructions teach AI how to read and interpret the legacy language: COBOL syntax and idioms, platform-specific patterns, data structure conventions, integration patterns. Specification generation instructions guide AI in how to structure specifications: what to include, what level of detail, how to document uncertainty, output format and templates. Domain-specific instructions provide context for your particular business area: benefits eligibility rules, tax calculations, licensing workflows, whatever applies to your system.

For your first week, assess what you need. What languages does your legacy system use? What platforms? What business domains? Then search for existing instructions in shared repositories or open-source collections. Other agencies practicing SpecOps, or something close to it, may have already created what you need, or something close enough to adapt.

In your second week, create basic instructions if you can't find existing ones to adapt. Keep them simple initially. Provide general guidance for the task. Include common patterns to look for. Add examples of the output you want. Document pitfalls to avoid based on what you know about your codebase.

Then, ongoing, refine these through repeated use. Every time AI produces poor output, treat it as an opportunity to improve the instructions. Capture what works and what doesn't. Build pattern libraries over time. And consider sharing your improvements back to the SpecOps community. The instruction sets you develop may help another agency facing similar challenges.

The First 90 Days

A realistic timeline helps manage expectations and ensures early progress builds on itself rather than creating chaos. Here's what the first three months typically look like for a SpecOps pilot.

Month one focuses on discovery, prioritization, and setup. In the first two weeks, assemble at least your core team, even if some positions remain unfilled. Set up the specification repository. Install and configure AI coding agents. Begin the system inventory and documentation review. Critically, initiate source code access requests now; they often take longer than expected. Also identify and request related materials like JCL, copybooks, schemas, and whatever documentation exists.

In weeks three and four, complete the component inventory. Verify that source code access is actually working, that you can extract and view the code, not just that someone approved your request. Identify domain experts and begin building those relationships. Assess risk factors and documentation state for each component. Begin drafting prioritization criteria. Catalog what materials you have versus what's missing.

By the end of month one, your deliverables should include: team assembled, tools configured, source code accessible, and system overview documented.

Month two focuses on pilot selection and initial specification generation. In weeks five and six, finalize your pilot component selection using the criteria discussed earlier. Get stakeholder buy-in on the choice. Define success criteria before you begin, not after. Prepare or select initial instruction sets for the legacy language and domain.

In weeks seven and eight, begin the actual SpecOps work. Feed pilot component source code to AI agents. Generate initial specifications. Have engineers conduct technical review of the AI-generated content. Refine instruction sets based on results; you'll learn a lot about what the instructions need in this phase.

By the end of month two, your deliverables should include: pilot component selected with success criteria defined, initial specifications drafted, and instruction sets refined based on experience.

Month three focuses on domain expert verification, the heart of what makes SpecOps different. In weeks nine and ten, schedule domain expert review sessions. Prepare supporting materials: examples, test cases, policy references. Identify specific questions and uncertainties you need resolved.

In weeks eleven and twelve, conduct active verification. Domain experts review specifications and identify what's correct, what's wrong, and what's missing. Capture their corrections and clarifications. Update specifications based on feedback. Begin documenting institutional knowledge that emerges during these sessions, knowledge that might otherwise have remained tacit.

By the end of month three, your deliverables should include: verified specifications for the pilot component and documented domain knowledge.[13]

After month three, the path continues: month four for implementation, month five for testing, month six for deployment of the first modernized component. Compare this to traditional approaches, where the same component might take six to twelve months, or to direct AI translation, which might be faster but carries much higher risk of errors.[14]

Timeline variations are normal. Simpler components move faster. Complex components or limited domain expert availability extend the timeline. What matters is that subsequent components get faster as instruction sets improve and patterns emerge. You're not just modernizing one component; you're building capability for modernizing many.

Managing Expectations

Stakeholder expectations must be actively managed. Early wins build confidence, but unrealistic promises destroy credibility. The political dimension of getting started is as important as the technical dimension.

When talking with leadership, be clear about what SpecOps can and can't deliver. You can promise knowledge preservation that creates lasting value, risk reduction through domain expert verification, incremental value delivery with results within months rather than years, clear audit trails through version-controlled specifications, and better outcomes than traditional approaches have produced.

You can't promise instant results, zero risk, that AI will do everything automatically, or that this eliminates the need for expertise. Anyone making those promises is setting up a team for failure.

The executive buy-in conversation has several components.[15] Risk reduction resonates with leaders who've seen or experienced modernization failures. Domain expert verification catches errors before implementation. The incremental approach reduces the big-bang failure risk that has sunk so many projects. Specifications provide the audit trail and accountability that government projects require.

Knowledge preservation resonates with leaders watching experienced staff retire. SpecOps captures institutional knowledge before it walks out the door. Specifications create a lasting asset that outlives any particular codebase. This reduces future modernization risk; the next time you need to modify or update a system, the specifications will still be there.

Value delivery resonates with leaders under pressure to show progress. Components deploy within months, not years. Return on investment comes early, not all at the end. Progress can be demonstrated continuously through completed specifications and verified components.

Frame the pilot as low commitment, high learning. It proves value with limited investment. It demonstrates the approach on an actual component. Success builds confidence for larger and sustained investment.

You'll encounter predictable objections. "This sounds like waterfall" is common because specifications suggest big-design-up-front. Explain that SpecOps is iterative; you specify, verify, implement, and deploy one component at a time. Work is deployed continuously. Specifications are living documents that evolve.

"We need to move faster" is understandable given the pressure leaders face. Show the risk of speed without verification. Remind them of projects that moved fast and failed expensively. Demonstrate that incremental delivery is actually faster to realized value than big-bang approaches that take years before delivering anything.

"Why not just translate the code directly?" questions the added effort of specification generation and verification. Explain the knowledge preservation value. Note that direct translation produces code almost nobody can verify is correct. Point to the long-term cost of "new legacy" created when translated code embeds old assumptions without anyone understanding them.

Early wins matter enormously for maintaining momentum and support. Celebrate when the first specification gets verified by a domain expert. Celebrate when AI correctly identifies business rules that would have taken humans days to untangle. Celebrate the first demonstration to stakeholders, the first reusable instruction set, the first specification that captures knowledge a retiring expert held. Each of these is concrete evidence that the approach is working.

Communicate progress regularly. Provide updates to leadership and stakeholders on a predictable schedule. Conduct early and frequent demos. Make specifications visible and accessible. Share lessons learned, both successes and challenges. Transparency builds trust.

Watch for common pitfalls in the early phase. Skipping domain expert verification to save time defeats the entire purpose of SpecOps; make verification non-negotiable from day one. Perfectionism, waiting for perfect specifications before proceeding, kills momentum; set "good enough" standards and iterate. Over-relying on AI, accepting AI-generated content without critical review, embeds errors; build in checkpoints that assume AI will make mistakes. Inadequate tooling, trying to execute without basic infrastructure in place, creates friction that slows everything; invest in the minimum viable toolchain before starting.

Beginning

The whiteboard in the Deputy CIO's conference room eventually filled with answers. The team selected an income verification module as their pilot. It had moderate complexity, clear boundaries, known behavior that their senior policy analyst could verify, and enough business value that stakeholders would notice success. They identified their core team: The Deputy CIO's technical lead would also serve as AI Engineering Lead initially, they'd bring in two engineers from a recent project, and their policy analyst would serve as Lead Domain Expert with support from a business analyst who knew the eligibility rules backward and forward.

They set up their specification repository that week. Getting source code access took two weeks longer than expected, a reminder that legacy systems live in organizational contexts as much as technical ones. Their first instruction sets were rough, but they improved with each iteration. By the end of month three, they had verified specifications for the pilot component, specifications that captured business rules their retiring mainframe programmer had never documented but that their policy analyst immediately recognized as correct.

They weren't done. Implementation, testing, and deployment still lay ahead. But they had demonstrated something important: the approach worked. They understood the legacy system's behavior in a way they never had before. They had captured knowledge that would otherwise have remained at risk. And they had done it in three months, not three years.

Getting started on any modernization project, particularly one that uses a new approach, is genuinely hard. There's organizational inertia to overcome, skepticism to address, and legitimate risk to manage. But SpecOps is designed to reduce risk through incremental progress. The pilot component is your proving ground. The team you assemble will learn and improve together. The toolchain will evolve as you use it. The first 90 days set the pattern for everything that follows.

Start small. Learn fast. Build confidence. Scale up. The hardest part is the first step. The next chapter will show you how to work effectively with the AI tools at the heart of this approach.

Chapter 14: Working with AI and People

The AI had generated a specification for the income verification module in under an hour. Two hundred lines of structured documentation describing calculations, thresholds, exception handling, and data flows. The technical lead was impressed. The output looked comprehensive, professionally formatted, and internally consistent. Then the policy analyst started reading.

"This isn't right," she said, pointing to a section on income exclusions. "We don't count child support as unearned income for this program. We did until 2019, but that changed." She scrolled down. "And this formula for the earned income disregard is backwards. It's supposed to be the first $200 plus half the remainder, not half the first $200."

The AI had produced something plausible but wrong. It had likely leveraged older documentation or conflated rules from similar programs. The errors weren't immediately obvious to anyone who didn't know the policy deeply. But they could have produced incorrect eligibility determinations for thousands of people if implemented as written.

This is the moment where SpecOps either works or doesn't. AI generates drafts quickly at scale. But only human expertise, specifically domain expertise, can verify whether those drafts are correct. SpecOps succeeds when it builds collaboration where each contributes what they do best: AI handling analysis and generation at scale, humans providing the judgment and domain knowledge that ensure accuracy.

Domain Experts at the Center

In most legacy modernization projects, domain experts participate at the edges. They're interviewed during requirements gathering at the beginning. They review documents when drafts are ready. They test the system during user acceptance testing at the end. But the core work of understanding the legacy system and building the new one happens without them, conducted by technical staff working from whatever documentation exists.

This model fails for predictable reasons. Requirements gathered early become stale as the project evolves. Documents reviewed late get approved without deep engagement because changing them at that point is too expensive. Testing catches some errors but misses the subtle policy misinterpretations that only deep domain knowledge would catch. The people who actually know how the business works are never truly integrated into the work that determines whether the new system will work correctly.

SpecOps takes a different approach. Domain expert verification sits at the center of the methodology, not the periphery. Every specification generated by AI goes through domain expert review before implementation begins. This isn't a quality gate tacked onto the end of a phase. It's the core activity that determines whether a specification is ready to proceed.

This changes what domain experts do. They're not approving documents that someone else created and handed to them for sign-off. They're actively verifying that the system's behavior has been correctly captured, examining specifications with the depth of attention that their expertise makes possible. They become the arbiters of correctness, and their judgment is authoritative. When a domain expert says a specification is wrong, that conclusion drives what happens next.

The approach aligns with principles that teams like 18F and USDS have emphasized for government digital service projects. The USDS Digital Services Playbook calls for cross-functional teams that include subject matter experts as full participants, working iteratively rather than in sequential phases.[1] It emphasizes keeping delivery teams small and focused, limiting organizational layers between teams and business owners, and using agile methods to reduce risk through frequent delivery and continuous feedback.[2] The partnership model once used by 18F embedded an empowered product owner from the partner agency who understood the organization and the problem being solved, working alongside the technical team throughout the project rather than just reviewing outputs.[3]

SpecOps applies these same principles to legacy modernization. As detailed in the previous chapter, domain experts are embedded in the team. Specifications are verified incrementally, component by component. Course corrections happen continuously rather than at the end. The methodology assumes that getting things right requires ongoing collaboration between people who understand the technology and people who understand the business, not handoffs between separate groups working in sequence.

Adjustments Required

Bringing domain experts into the center of the work requires adjustments from everyone involved.

Domain experts need to engage with technical artifacts in ways they may not have before. Specifications are more detailed and precise than policy memos or procedure manuals. They describe system behavior explicitly, including edge cases and exception handling that might never appear in higher-level documentation. Domain experts don't need to become programmers, but they do need to read specifications carefully and evaluate whether the described behavior matches their understanding of how the system should work.

This often requires training and support. Domain experts may need to learn how specifications are structured, what level of detail to expect, and how to provide feedback that technical staff can act on.[4] A comment like "this section seems wrong" is less useful than "this formula applies the disregard before the exclusion, but it should be the other way around." The investment in building this capability pays off because it makes verification meaningful rather than ceremonial.

Technical staff, for their part, need to genuinely value domain expertise. When a domain expert says a specification is wrong, that judgment should be taken seriously, investigated thoroughly, and resolved before proceeding. Dismissing domain expert concerns, treating verification as a hurdle to get past rather than the checkpoint where accuracy gets established, undermines the entire methodology.[5]

This also means writing specifications that domain experts can actually engage with. Specifications so technical that only engineers can read them defeats the purpose. The skill of translating system behavior into language that's precise enough to be useful but accessible enough for non-technical reviewers becomes essential. Business analysts often play a crucial bridging role here, helping structure specifications so that both technical staff and domain experts can work with them productively.

Building Trust

Effective collaboration between technical staff and domain experts requires mutual trust, and that trust develops over time through demonstrated competence and respect.

Trust runs in both directions. Domain experts need to trust that engineers are accurately capturing system behavior in specifications and will take verification feedback seriously. Engineers need to trust that domain experts are verifying thoroughly and catching real errors, not just skimming and approving. Both groups need to believe the other brings essential expertise that they themselves lack.

This trust doesn't exist automatically, especially when people are working together for the first time or when the collaboration model is new to the organization. It develops through working together, seeing each other's contributions, and experiencing the value of the collaboration directly.

Early wins build this trust. The first time a domain expert catches a significant error, like the income exclusion mistake in the opening example, engineers see concretely why verification matters. They see that domain expertise catches things they could have missed, things that could have caused real problems if implemented. The first time AI correctly extracts complex business rules that would have taken humans days to untangle, domain experts see the value of the technical approach. They see that the AI tools aren't just generating noise but are doing useful work that makes their verification contribution valuable.

Cross-functional teams that work together daily build trust faster than groups that interact only at handoff points. When technical staff and domain experts are in the same meetings, reviewing the same artifacts, and solving problems together, they develop working relationships that support honest feedback and productive disagreement. Someone who has worked alongside you for weeks will tell you directly when something is wrong. Someone who only sees your work at formal review checkpoints may be more hesitant.

Signs of Healthy Collaboration

How do you know if the collaboration between technical staff and domain experts is working? Look for these indicators:

Indicator	Description
Domain experts feel comfortable saying "this is wrong"	They raise concerns without hesitation, confident their input will be taken seriously.
Technical staff investigate domain expert concerns thoroughly	Verification feedback drives real changes, not defensive pushback.
Both groups can explain why the other's expertise matters	Engineers articulate why domain verification catches errors they'd miss. Domain experts articulate why AI-assisted analysis makes their verification time worthwhile.
Disagreements get resolved through evidence and discussion	When technical and domain perspectives conflict, the team investigates rather than defaulting to one side.
Early errors are celebrated as catches, not blamed as failures	Finding a mistake before implementation is a success for the process, not an indictment of whoever made it.

If these indicators are absent, address the underlying dynamics before they undermine the methodology.

Trust erodes when concerns are dismissed, when expertise is devalued, or when one group believes the other isn't taking the work seriously. Technical staff who push back on every domain expert correction, or who treat verification as a formality to rush through, will find domain experts disengaging. Why invest time in careful review if the feedback gets ignored? Domain experts who refuse to engage with technical details, or who approve specifications without reading them carefully, will find technical staff losing confidence in the verification process. Why generate detailed specifications if no one actually checks them?

Watch for these warning signs and address them directly. The collaboration this methodology requires can't survive sustained distrust between the groups whose cooperation makes it work.

Working with AI

AI tools generate drafts quickly but can make confident errors. Using them effectively requires understanding what they do well, where they can fail, and how to structure work to get the benefits while catching the mistakes.

AI excels at processing large volumes of legacy code and generating structured output. It can read thousands of lines of COBOL and produce a draft specification in hours rather than months. It can identify patterns, extract business rules, and organize complex logic into readable documentation. This is work that would take humans far longer and would depend on scarce legacy expertise that may not be available. For organizations facing the retirement of their last COBOL programmers, AI's ability to analyze legacy code is not just convenient but essential.

AI also improves with guidance. Instruction sets that provide context about the system, the domain, terminology, and the desired output format produce better results than generic prompts. Part of the AI Engineering Lead's job is refining these instruction sets based on what works and what doesn't, building institutional knowledge about how to get good output from AI tools for your specific systems and domains.[6]

But AI does make errors that can look plausible. It may confuse similar programs, mix up policy versions from different time periods, or generate calculations that are internally consistent but factually wrong. It hallucinates details when it lacks information, filling gaps with reasonable-sounding content that happens to be incorrect. It doesn't know what it doesn't know, and it doesn't always flag uncertainty in ways that make errors obvious.

These errors are often subtle. The income exclusion example from the opening is typical: someone without deep domain knowledge wouldn't notice the problem. The specification reads well. The logic flows. It's only wrong if you know what right looks like. This is why domain expert verification isn't optional. It's why verification is where accuracy actually gets established.

Common AI Error Patterns

AI-generated specifications fail in characteristic ways. Knowing these patterns helps domain experts focus their verification:

Error Pattern	Description
Confusing similar programs or policy versions	AI may conflate rules from related but distinct programs, or mix requirements from different benefit categories.
Using outdated rules that have since changed	Training data includes historical documentation. Policy changes from recent years may not be reflected, or old and new rules may be blended incorrectly.

Error Pattern	Description
Generating plausible calculations that are factually wrong	Formulas may be internally consistent and mathematically valid while applying the wrong logic entirely.
Filling gaps with hallucinated details	When source material is incomplete, AI generates reasonable-sounding content rather than flagging uncertainty. These invented details look authoritative.
Producing internally consistent but externally incorrect specifications	The document reads well and the logic flows. It's only wrong if you know what right looks like.

Domain experts should read specifications asking "what might be wrong here?" rather than "does this look okay?" These error patterns suggest where to look most carefully.

The verification process should assume AI has made mistakes and look for them actively. Domain experts should read specifications with the question "what might be wrong here?" rather than "does this look okay" Critical reading catches errors that casual review misses. The goal isn't to distrust AI but to verify AI, which is a different posture than simply accepting output because it looks reasonable.[7]

Teams need to calibrate their trust in AI output appropriately over time. Over-trusting AI leads to implementing plausible errors. Under-trusting AI leads to not using it effectively, losing the speed and scale benefits that make the methodology practical. The right calibration comes from experience: seeing where AI performs well, learning what kinds of errors it tends to make, and adjusting review intensity accordingly.

This calibration develops as the team gains experience with specific instruction sets and specific types of specifications. Early in a project, more intensive verification makes sense while everyone is learning AI's patterns. As the team develops intuition about where to look most carefully, verification can become more targeted. But it never becomes unnecessary. Even experienced teams using mature instruction sets will encounter AI errors that only domain expertise can catch.

Two Kinds of Change at Once

SpecOps requires both adopting new tools and changing how people work together. Managing these simultaneously is harder than managing either alone.

Learning new AI tools is one kind of challenge. The tools have capabilities and limitations that take time to understand. Prompt engineering, instruction set design, and interpreting AI output are skills that develop with practice. People who have never worked with AI coding assistants need time to build proficiency.

Changing collaboration patterns is a different kind of challenge. Bringing domain experts into the core work loop, building trust between technical and domain staff, establishing verification as a central activity rather than an end-stage checkpoint: these are organizational changes that affect relationships, roles, and expectations. They can encounter resistance from people comfortable with existing ways of working.

Doing both at once creates compounding difficulty. Someone struggling with the AI tools might attribute their frustration to the new collaboration model. Someone uncomfortable with the changed role of domain experts might blame problems on the technology. Separating these dynamics helps address them. When something goes wrong, ask: is this a tool problem or a collaboration problem? Different problems need different solutions.

Some teams benefit from establishing AI tool proficiency before changing collaboration patterns significantly. Get a smaller group comfortable with the technology, develop working instruction sets, understand what good AI output looks like. Then expand the collaboration to include domain experts in verification. Other teams find that the new collaboration model is what motivates learning the tools: domain experts see the value of verification, that engagement drives investment in making the AI tools work well, and capability builds across both dimensions together.

There's no universal right answer. Consider your team's starting point. If AI tools are entirely new to everyone, some ramp-up time with a smaller group may help avoid overwhelming people with too much unfamiliarity at once. If the collaboration model is the bigger change because domain experts have historically been excluded from technical work, start with the full cross-functional team and learn the tools together.

Either way, both changes require experimentation, and experimentation means mistakes. Teams need permission to try things, fail, learn, and adjust without those failures becoming career risks or evidence that the approach doesn't work.

Share problems openly. When AI generates something wrong, discuss it as a team so everyone learns what to watch for. When collaboration breaks down, address it directly rather than letting frustration build. Gene Kim and Steve Yegge, writing about organizational adoption of AI-assisted development, emphasize normalizing the learning process: share the disaster stories, make clear that mistakes are expected, create conditions where people are excited to experiment knowing there will be missteps along the way.[8] The same principle applies here.

SpecOps is new enough that every team adopting it is learning. Treat that learning as part of the work, not as evidence of failure.

Making It Work

The policy analyst's corrections took twenty minutes to document. She walked the technical lead through both errors, explaining not just what was wrong but why: the history behind the 2019 policy change, the logic of the earned income disregard formula, the upstream and downstream effects of getting these calculations wrong.

The technical lead updated the instruction set to include explicit guidance about the 2019 policy change and added a note about the correct disregard formula structure. When they regenerated the specification with the improved instructions, both errors were fixed. More importantly, similar errors were less likely to appear in future specifications for related components.

The team had demonstrated the pattern they would use throughout the project. AI generated drafts quickly. Domain experts verified them carefully. Errors got caught before they could cause harm. When verification identified problems, the fixes improved not just the immediate specification but the instruction sets that would guide future work. The errors that were identified became data to enhance the instruction sets used for the next cycle of the process. The collaboration worked because everyone understood their role and trusted each other to perform it.

This pattern doesn't emerge automatically. It requires domain experts who engage as full participants, not peripheral reviewers consulted only at milestones. It requires technical staff who genuinely value domain expertise and respond to verification feedback as authoritative rather than advisory. It requires AI tools used with appropriate trust: confident in their ability to generate useful drafts, but skeptical enough to verify everything before implementation. And it requires organizational conditions that support learning, where mistakes are treated as information rather than failures.

The next chapter addresses what happens when this collaboration faces obstacles. Domain experts may not be available. Legacy code may be genuinely incomprehensible. Undocumented features may emerge that no one expected. The foundation built through effective collaboration determines how well a team can handle these challenges when they arise.

Chapter 15: Common Challenges and Solutions

The project manager stared at the org chart with growing concern. Of the four people who intimately understood the benefits calculation module, one retired eight weeks ago, one was retiring in six weeks, one had transferred to another agency, and one was on extended medical leave. The modernization initiative had finally gotten funding and approval. The window for capturing institutional knowledge was closing faster than the project could ramp up.

This scenario plays out in a similar fashion across government. The same workforce dynamics that make legacy modernization urgent also make it harder to execute. The people who understand these systems are leaving. The expertise needed to verify specifications is literally walking out the door. An approach that depends on domain expert verification faces a real challenge when domain experts are increasingly scarce.

This isn't a reason to abandon SpecOps. It's a reason to adapt it. The methodology's emphasis on capturing knowledge in specifications becomes more valuable, not less, when expertise is disappearing. But teams need practical strategies for situations where ideal conditions don't exist.

When Domain Experts Aren't Available

The book has established that government at all levels faces a workforce crisis. COBOL programmers average between 50 and 70 years old. Ten thousand Baby Boomers retire every day. Federal IT staff departures are accelerating. The people who understand legacy systems are the same people leaving the workforce.

This creates a paradox for SpecOps. The methodology depends on domain expert verification. But domain experts are increasingly unavailable. But if we wait for perfect conditions, we'll wait until the expertise has vanished entirely. The only option is to adapt.

The aging workforce is a problem for any legacy modernization approach. Traditional approaches take so long that they typically can't retain expertise throughout. A five-year waterfall project that starts with requirements gathering will finish long after the domain experts who could verify those requirements have retired. The knowledge captured in year one becomes stale, unverifiable, and disconnected from the people who once understood the system.

SpecOps addresses this through iterative, component-by-component knowledge capture. Rather than attempting to document everything about a system upfront, teams work through components incrementally. Each component's specification is created and verified while moving toward implementation. The urgency is real, but the appropriate response is prioritization: identify which components face the greatest knowledge-loss risk and work on those first, capturing expertise iteratively as you go.

Specifications created and verified for each component remain valuable even after the experts who verified them have left. The specification becomes institutional memory that persists. But this capture happens as part of the ongoing modernization work, not as a separate, up front documentation phase.

When expertise is at risk, prioritize the components that expertise covers. A domain expert six months from retirement can't participate in modernizing every component. But they can be involved in specifying and verifying the components they know best while those components move through the SpecOps methodology. If the person who understands the benefits calculation module is retiring in six months, that module moves up the priority list.[1]

When the ideal domain expert isn't available, teams need to identify alternative sources of knowledge. Frontline staff who process applications may not know why the system works the way it does, but they know what it does and where it fails. Reports generated by the system encode expectations about its behavior. Regulations and policy documents describe intended behavior even if actual implementation has drifted. Change logs and ticket histories can provide context about why systems behave the way they do. Other agencies with similar systems may have developed instruction sets that can inform your work.[2]

None of these fully substitute for deep domain expertise. But they can reduce the verification burden and fill gaps when ideal expertise isn't available.

Sometimes verification proceeds without the expertise needed for full confidence. This reality must be acknowledged and managed. Denote specifications with confidence levels: "verified by domain expert with 20 years experience" is different from "verified against policy documents by analyst new to the program." Both can proceed, but downstream users need to know the difference. Prioritize high-risk areas for expert verification. Not every specification needs the same level of scrutiny. Calculations that determine benefit amounts warrant more verification investment than logging routines.[3]

The window for capturing institutional knowledge is closing. Every month of delay means more expertise lost. The response isn't to abandon iterative development for a waterfall-style documentation phase. The response is to start now and prioritize strategically. Begin working through components, sequencing based on knowledge-loss risk as well as business value. Each component that moves through the methodology captures knowledge in durable form. Waiting for perfect conditions means potentially waiting until expertise has vanished entirely.

Handling Legacy Code That's Truly Incomprehensible

Some legacy code defies understanding. Decades of patches, lost documentation, departed expertise, and organic evolution create systems that no one fully comprehends. Not all incomprehensible code is equally opaque. Some code is structurally complex but logically coherent: tangled, but patient analysis can extract some logic. Some code is poorly documented but behaviorally observable: no one knows why the code works, but testing can reveal what it does. Some code is truly orphaned: no documentation, no experts, behavior that varies unpredictably, or code that seems to contradict itself.

When understanding the code directly fails, shift to understanding its behavior. Run the system with test data and observe outputs. Trace data flows through interfaces. Identify patterns even if you can't explain their origin.

AI tools can help, especially when guided by well-developed instruction sets. A team struggling to understand a COBOL benefits calculation module may find that instruction sets developed by other organizations working with similar systems provide the context AI needs to make sense of the code. The instruction set encodes hard-won knowledge about COBOL idioms, common patterns in benefits systems, and domain-specific terminology that helps AI produce more useful analysis.

This is where the shareable nature of SpecOps artifacts pays off. As discussed in Chapter 12, instruction sets are highly portable. COBOL is COBOL everywhere. The patterns in one state's unemployment system may resemble patterns in another's.

An instruction set refined through months of work at one agency can help another agency's team make progress on code that would otherwise defy understanding. When your own analysis stalls, look for instruction sets from other organizations working with similar legacy platforms. The code may be incomprehensible to you, but someone else may have already taught AI how to make some sense of it.[4]

Chapter 14 discussed AI's tendency to produce plausible-but-wrong output. This risk is amplified when working with incomprehensible code. Normally, engineers can verify AI analysis against their own understanding. When no one understands the code, that check disappears. AI can confidently describe what code does, and no human can say "that's not right" from direct knowledge.

Confidence Levels for Specifications

Level	Description	Verification Basis
1	Verified by domain expert with deep system knowledge	Expert with 10+ years experience reviewed and approved
2	Verified against policy documents by program analyst	Analyst confirmed alignment with current regulations and policy
3	Verified by observation of system behavior only	Behavioral testing confirms outputs match specification
4	Documented but not yet verified	AI-generated, pending review

This demands additional verification strategies. Use multiple AI models and compare their analyses. Where they agree, confidence increases. Where they disagree, investigate further. Neither may be right, but disagreement surfaces uncertainty that a single model's confident answer might hide. Validate AI claims against observed behavior. If AI says the code calculates income by summing monthly values, run test cases and confirm the outputs match that logic. Behavioral testing becomes the primary verification mechanism when code review fails. Look for internal contradictions in AI-generated specifications, reading critically for statements that can't both be true.[5]

Document the verification basis for each specification section. "Verified by domain expert review" is different from "consistent with behavioral testing" is different from "AI analysis only, not independently verified." Downstream users need to know which parts rest on solid ground and which represent best-effort analysis of code no one fully understands.

When underlying logic can't be understood, specifications can document observed behavior without claiming to understand it. "When input X falls between these values, output Y is calculated using this formula" is a valid specification even if no one knows why that formula was chosen. This approach has limitations: it may perpetuate bugs or inefficiencies, and it can't distinguish intended behavior from accidents. But it provides a foundation that pure code translation can't: a human-readable description of behavior that can be examined, questioned, and eventually understood.

Some code can't be understood with available resources. Teams need permission to acknowledge this. Denote components as "black box" with documented behavior but unknown logic. Implement based on observed behavior with explicit acknowledgment of risk. The worst outcome is pretending to understand what you don't. Specifications that claim comprehension where none exists create false confidence. Honest acknowledgment of uncertainty allows appropriate risk management.

Dealing with Undocumented Features and Workarounds

The code is only part of the legacy system. Around every legacy system is a web of human processes, manual interventions, and workarounds that compensate for system limitations. Caseworkers who know to double-check certain calculations manually. Batch operators who restart jobs in a specific sequence when they fail. Users who avoid certain input combinations because "the system does weird things." This knowledge is often undocumented and exists only in practice. When the system is modernized, these workarounds may be forgotten unless deliberately captured.

Workarounds emerge from multiple sources. Direct observation of users interacting with the system reveals behaviors that code review might never surface. Interviews that ask "what do you do when the system doesn't work right?" uncover compensating processes and steps. Error logs and ticket histories may reveal recurring problems that users have learned to work around. Training materials, especially informal ones, often document behaviors that formal documentation ignores. Look for gaps between documented process and actual process. Where do users deviate from official procedures? Why? What would happen if they followed the documentation exactly?[6]

Not every workaround should be perpetuated in the modern system. Some compensate for bugs that should be fixed. Some reflect outdated requirements that have changed. Some are genuinely necessary adaptations that the modern system must support. The specification process forces these decisions. Document what users actually do. Flag it as workaround behavior. Decide whether the modern system should require the same workaround or fix the underlying issue. This is where domain expert judgment is essential. Engineers can document workarounds. Only domain experts can determine whether they should persist.

When workarounds must be kept, specifications should document them explicitly, not as mysterious requirements but as acknowledged adaptations. "The system calculates preliminary eligibility, but caseworkers manually verify income documentation for applications over $50,000 because the automated verification occasionally misses employer-reported discrepancies." This captures the behavior, explains the rationale, and makes visible what was previously invisible. Future teams can decide whether to automate the workaround, fix the underlying issue, or continue the manual process.

When Specifications Reveal Bugs vs. Features

Specification work will surface discrepancies between actual system behavior and intended behavior. AI generates a specification. A domain expert reviews it. Something doesn't look right. The specification accurately describes what the code does, but what the code does isn't what policy actually requires.

This happens regularly. Legacy systems accumulate bugs that become embedded in operations. Workarounds develop around them. Changing the behavior would change outcomes for real people.

Not all discrepancies are clear-cut bugs. Some behaviors were intentional when implemented but policy has since changed. Some were never correct but have become de facto policy through longevity. Some are clearly wrong but fixing them would harm people who have come to depend on the incorrect behavior. The spectrum runs from obvious bugs through ambiguous cases to embedded behaviors that are incorrect but operationally depended upon.[7]

When specifications reveal discrepancies, teams need a framework for deciding how to proceed. Document both behaviors: what does the code do, and what should it do according to current policy? Make the gap explicit. Assess impact: how many people are affected, and what are the consequences of changing versus preserving the behavior? Escalate appropriately: some decisions can be made by the project team, while others require policy leadership, legal review, or executive decision-making. Document the decision: whatever is decided, record the reasoning so future teams understand why the modern system behaves as it does.[8]

Fix bugs when the incorrect behavior clearly violates current policy, the impact of changing is manageable, legal and policy authorities approve the correction, and affected parties can be notified and accommodated. Modernization creates an opportunity to fix accumulated errors. Take this opportunity when it is appropriate to do so.

Preserve what is identified as incorrect behavior when making changes would cause significant harm to people who depend on that behavior, legal or regulatory constraints prevent correction, the political or operational cost of changing exceeds the benefit, or maintaining compatibility during transition requires consistent behavior. Preserving bugs is a legitimate choice when deliberately made and documented. The specification should note that the behavior is known to be incorrect and explain why it's being preserved.

The value of specifications is that they make these decisions explicit. Whether you fix the bug or preserve it, the specification documents what the system does and why. Future teams inherit understanding, not mystery.

Balancing Speed and Thoroughness

Budget cycles create deadlines. Political windows are narrow. Domain experts retire. Stakeholders want results. The pressure to skip steps, abbreviate verification, and declare specifications "good enough" may be constant.

Understanding this pressure as legitimate doesn't mean capitulating to it. The legacy modernization failures catalogued in Part I resulted from giving in to schedule pressure when the work wasn't ready. But ignoring schedule pressure entirely can result in projects that never finish.

Verification catches errors before they propagate. Errors caught during specification verification are cheap to fix: update the specification, then regenerate. Errors caught during testing are more expensive. Errors caught in production are very expensive and potentially harmful. Time invested in verification is repaid in errors avoided down the road.[9]

Skipping verification doesn't eliminate errors. It delays their discovery. Abbreviated specifications that omit edge cases produce implementations that fail on those edge cases. Rushing through domain expert review produces rubber-stamp approval that catches little if anything. The verification checkpoint exists to catch errors. Going through the motions while actually checking nothing provides zero benefit.

Some efficiencies are genuinely available without sacrificing quality. Better instruction sets can produce higher-quality first drafts, reducing revision cycles. Experienced teams will work faster the second time through the process. Sequencing work to build capability on manageable components before tackling the hardest ones produces faster overall progress. Calibrating thoroughness to risk avoids wasting effort on low-risk components.[10]

When stakeholders demand faster delivery, the response shouldn't be a straight "no." It should be a conversation about tradeoffs. "We can deliver faster if we reduce scope. Which components are essential for the first phase?" "We can skip verification on low-risk components but not on benefit calculations. Here's what that revised timeline looks like." Make tradeoffs explicit. Let decision-makers understand what faster delivery actually requires.

Some steps can't be shortened without fundamental compromise. Domain expert verification can be more intensely focused on highest-risk areas, but eliminating it defeats the methodology's purpose. Specification accuracy can't be traded away. Testing against specifications can't be abbreviated below the level needed to confirm implementation correctness. These are the load-bearing walls. Everything else can be negotiated.

Making Progress in Imperfect Conditions

The project manager adjusted the plan. The benefits calculation module, which the retiring expert knew best, moved to the top of the component priority list. Rather than trying to document everything she knew before leaving, the team worked through the methodology on that module while she was still available: specification, verification, implementation. Video-recorded walkthroughs captured context for components that wouldn't be taken up before her departure. The recently retired expert agreed to consulting sessions for verification on future components. Users who processed applications daily were brought in to help identify the workarounds that existed nowhere in documentation.

Working iteratively but strategically, the team captured more institutional knowledge than the agency had documented in thirty years. Not by front-loading a massive documentation effort, but by prioritizing which components to tackle based on which set of expertise was most at risk.

The early work paid dividends beyond the components directly addressed. The team developed experience with the new methodology that made work on subsequent components faster. The instruction sets refined while the domain expert was still available improved AI analysis of related components tackled later. Patterns recognized in the first modules informed understanding of later ones. The specifications themselves provided context that helped interpret code in adjacent areas. Early investment in learning the system created leverage for the work that followed.

This is the reality of legacy modernization. Conditions are rarely ideal. Experts retire, code confounds analysis, workarounds hide in plain sight, bugs masquerade as features, and stakeholders demand impossible timelines. The challenges are real.

But they're not reasons to abandon the effort. They're reasons to adapt, to prioritize ruthlessly, to capture what can be captured while the window remains open, and to proceed with honest acknowledgment of uncertainty where perfect knowledge isn't available.

The methodology survives contact with reality not by ignoring these challenges but by providing strategies for addressing them. Specifications remain the source of truth even when verification is imperfect. Domain expertise remains essential even when experts are scarce. Knowledge preservation remains valuable even when complete understanding is impossible.

Part VI looks to the future, to breaking the legacy cycle and building systems that don't become the next generation's modernization crisis, and the broader implications of specification-driven development. The practical challenges addressed in this chapter inform that forward-looking vision. Sustainable government technology requires not just better methods but realistic expectations about the conditions under which those methods must operate.

Part VI

Looking Forward

"The best is yet to do, and here, where you are, they are coming to perform it"

As You Like It; Act 1, Scene 2

Chapter 16: Breaking the Legacy Cycle

In April 2020, New Jersey Governor Phil Murphy stood before cameras pleading for COBOL programmers. His state's unemployment system, built on decades-old technology, was collapsing under the strain of pandemic-era claims. It was a moment that crystallized the challenges of legacy technology systems in government: a fragile, poorly understood system, maintained by a shrinking pool of experts, untouched for so long that changing it had become difficult and risky.

New Jersey's crisis wasn't about old code—all code ages. It was about code that hadn't meaningfully evolved in decades. Somewhere along the way, the system had become untouchable. Changes were too risky because no one fully understood what the system did. Documentation had drifted from reality. The experts who could have guided modifications over time had retired. Each year of paralysis made the next change harder, and riskier, until the system existed in a kind of frozen state: running, but incapable of adaptation.

Now imagine a different history. Imagine a state that had treated its unemployment system as a product requiring continuous investment rather than a project completed decades ago. A state where specifications captured what the system did and were updated whenever the system changed. Where those specifications enabled small, verified modifications year after year, each one low-risk because the behavior was documented and domain experts could confirm correctness. Where the system in 2020 wasn't running decades-old technology because it had evolved incrementally over the decades it had been in use, shedding obsolete components and adopting new ones without requiring a crisis-driven overhaul.

In that scenario, the pandemic still creates unprecedented demand. Claims still surge. But the system isn't brittle because it hasn't been frozen in time. The team responding to the crisis has been making changes to it all along. They understand how the system works because they've been working with verified specifications continuously. The surge is a stress test, not a breaking point.

This isn't a fantasy about systems that never need attention. It's a vision of critical technology systems that receive attention continuously, in small increments, guided by preserved knowledge in specifications. What separates New Jersey's actual crisis from this alternative one isn't just the presence or absence of specifications—it's whether those specifications are used to enable the ongoing evolution that keeps technology systems from becoming untouchable.

The previous chapters in this book described an ongoing crisis, introduced a new approach to address it, detailed why this approach can work for government, and laid out practical guidance for implementing it. This chapter explains what success can actually look like.

What would it mean to build government technology that evolves continuously rather than remaining largely untouched while expertise retires until crisis forces change?

Breaking the Cycle

Before we can break the legacy cycle, we need to name it clearly. This cycle has repeated for decades because the same structural dynamics keep producing the same outcomes.

Chapter 2 traced how government technology reached its current state: mainframe investments that made sense in the 1960s, budget structures optimized for physical infrastructure, procurement systems that favor large contracts, and organizational dynamics that accumulate technical debt. These forces didn't stop operating after the first generation of legacy systems. They're still operating today.

The code being written today will eventually become legacy code. The "modern" systems replacing COBOL will themselves need replacement someday. The "modern"" applications deployed this year will eventually seem as dated as COBOL does now. What's not yet clear is whether the next transition away from these systems will be as painful and disruptive as the current transition.

Breaking the Cycle: Key Themes from This Book

Insight	Chapters	Implications
Knowledge is the valuable artifact	Chapters 5, 7, 9, 10	Preserve specifications, not just code
Small over large	Chapters 2, 3, 8, 11	Incremental evolution, not big-bang replacement
Process shapes outcomes	Chapters 2, 11	Change incentives, not just technology
Domain expertise + AI	Chapters 4, 7, 8, 9, 14	New division of labor between humans and tools
Technology choices constrain successors	Chapters 1, 2, 6	Design for the next modernization
Collaboration multiplies value	Chapter 12	Share instruction sets, not just code

Traditional modernization efforts focused exclusively on system replacement won't break this cycle. They will simply reset the countdown to crisis. New systems will inherit the same structural vulnerabilities: knowledge concentrated in the people who built them, documentation that drifts from reality, business logic embedded in implementation details that future teams will struggle to understand. The Phoenix pay system, examined in Chapter 3, replaced a 40-year-old legacy system with modern technology. Within years, it was generating its own mountain of technical debt, its own documentation gaps, its own institutional knowledge problems.[1] Modernization alone doesn't prevent the next legacy crisis. It just determines when it arrives.

Breaking the cycle requires changing what we preserve. When code is the only authoritative record of system behavior, knowledge is locked in an artifact that becomes progressively harder to understand. Languages fall out of fashion. Frameworks become unsupported. The people who wrote the code eventually move on. Each year, the code becomes more opaque, until changing it safely requires expertise that may no longer exist.

When specifications are the source of truth, knowledge is preserved in a form that persists regardless of implementation technology. A specification describing eligibility rules doesn't become unreadable because the underlying code migrated from Java to Node or from on-premises servers to cloud infrastructure. The business logic is the same. The specification captures it in terms that domain experts can verify today and successors can understand decades from now.

This is the fundamental insight that emerged across Part II and Part III of this book: the specification is more valuable than the code.[2] Code is one possible implementation of a specification. Specifications can guide many implementations across many technologies. If we preserve specifications and treat code as derivative, we break the pattern of knowledge loss that makes each modernization a crisis.

Designing for the Next Modernization

Every system built today will eventually need modernization. Designing with that reality in mind changes what we build and how we build it.

Twenty years ago, the iPhone didn't exist. Cloud computing was nascent. Social media was emerging. The technology landscape was unrecognizably different from today. Twenty years from now will be equally different in ways we can't predict. The specific technologies we choose today will become dated. The languages and frameworks that seem modern now will seem quaint. But the specifications that describe what systems do don't have to age the same way. A well-written specification transcends the implementation choices of any particular era.

When done well, traditional approaches to system design focus on making technology choices that will evolve gracefully over time, and be easy to maintain. But predicting the future is hard. Making government technology more sustainable is not just about trying to make the best choices today, it's about preserving knowledge in forms that transcend implementation choices entirely.

This means treating specifications as the authoritative source of knowledge, as Chapter 9 established. The specification describes what the system does. The code implements the specification. When there's a discrepancy, the specification is what we trust; the code is what we fix. This isn't just a philosophical stance. It's an operational discipline that determines whether knowledge persists or evaporates. Whether systems adapt or are frozen.

It means ensuring that changes flow through specifications, as the SpecOps methodology requires.[3] Every modification starts with a specification update. This creates an audit trail, ensures appropriate review, and keeps documentation synchronized with reality. The "we'll document it later" pattern that Chapter 3 identified as a consistent failure mode becomes structurally impossible when the specification must change before the code can.

It means maintaining domain expert verification at every stage, as Chapters 8 and 14 detailed. The people who understand policy verify that specifications correctly describe intended behavior. This catches errors early, but more importantly, it captures institutional knowledge continuously. Each verification session is an opportunity to document understanding that would otherwise evaporate.

And it means embracing incremental evolution rather than periodic crisis. The Strangler Fig pattern—and other incremental replacement patterns discussed in Chapter 8—enables systems to change continuously in small increments rather than accumulating change until a crisis forces massive replacement.[4] Each increment is low-risk because the scope is limited and the behavior is specified. The system never becomes frozen because it never stops evolving.

When a domain expert retires, their specifications they have helped verify remain. When a new team takes over, they can read what their predecessors understood. When auditors need to know how a system implements policy, they can trace from statute to specification to code. The knowledge doesn't disappear with personnel changes. It persists in artifacts designed to outlast any individual's tenure.

The Compounding Value of Preserved Knowledge

Knowledge preservation isn't just about avoiding the next legacy system crisis. Preserved knowledge compounds in value over time, enabling capabilities that weren't possible before.

Each specification verified is organizational knowledge captured. Each component modernized using the SpecOps methodology leaves behind documentation that supports future work. The instruction sets refined while working on early components improve AI analysis of later ones. Patterns recognized in first modules inform understanding of related systems. The investment in knowledge capture pays future dividends.

An organization that practices specification-driven development for five years accumulates a library of verified specifications covering its core systems. That library becomes an asset with multiple uses: training material for new staff who can learn systems by reading specifications rather than reverse-engineering code, reference documentation for auditors who need to understand how systems implement policy, foundation for future modernization efforts that can start from documented understanding rather than archaeological excavation, and institutional memory that doesn't retire when individuals do.

Chapter 12 discussed why government code sharing has fallen short of expectations: policy logic varies by jurisdiction, technology stacks and in-house capabilities are different, and the economics of software development can discourage building for reuse.[5] But instruction sets operate at a different layer. COBOL is COBOL whether you're modernizing benefits in California or taxes in New York. Legacy platform patterns are similar across governments even when business rules differ.

As more agencies develop and share instruction sets, the collective value grows. The work one state does to teach AI assistants how to comprehend mainframe batch processing helps every other state facing similar systems. The domain patterns one agency identifies in benefits eligibility calculations inform similar work elsewhere. The shared library of instruction sets becomes a public good that accelerates modernization across government, even though the actual systems and policies remain jurisdiction-specific.

Early contributors to shared instruction libraries create value for everyone who follows. The accumulated knowledge from dozens of modernization efforts becomes available to any agency starting new work. The isolation that has characterized government technology, every agency solving the same problems independently, begins to break down. Not through shared code, but through shared knowledge about how to understand and work with common legacy platforms.

Chapter 11 described the accountability requirements unique to government: systems must be auditable, explainable, and traceable in ways that code by itself often can't support.[6] Version-controlled specifications address this directly. They show what changed, when, and why. Decision records document the rationale for choices. Domain expert sign-off creates clear accountability for verification. The specification repository becomes an institutional record that supports the oversight democratic government requires.

When legislators ask how a system implements a particular policy, the answer isn't "it's complicated" or "you'd have to read the code." The answer is a specification that traces from statute to business rules to system behavior, written in language that policy staff can understand. When auditors question whether a system correctly implements eligibility requirements, they can review specifications rather than attempting to interpret source code. The accountability gap that makes government technology oversight so difficult begins to close.

What This Means for How We Work

Breaking the legacy cycle requires changes not just to technology but to how government approaches technology work. The methodology has implications for teams, budgets, and organizational structures.

Chapter 2 described how budget structures usually treat IT systems as projects with beginnings, middles, and ends rather than products requiring sustained investment.[7] You fund a capital project to build the system, then you fund a smaller operations and maintenance budget to keep it running. The project ends, the development team disperses, and the system enters what's euphemistically called "maintenance mode," which in practice often means managed evolution into obsolescence.

This model guarantees the legacy cycle. A system that stops receiving development investment starts accumulating technical debt from day one. Documentation drifts. Knowledge concentrates in whoever happens to be maintaining it. The gap between what the system does and what anyone understands about it widens with each passing year. Eventually the system becomes untouchable, and the cycle completes.

Breaking the cycle requires treating systems as products: continuously developed, continuously maintained, continuously improved. This is a budget process change, not a technology change. But it's a change that specification-driven development makes more visible and more achievable. When the specification is the source of truth, the ongoing investment in keeping specifications current and verified is an explicit line item, not a hidden cost buried in maintenance budgets. The work of preserving knowledge becomes visible work that can be planned, funded, and measured.

The skills that matter shift as well. Chapter 4 described how AI coding assistants change what expertise is most valuable. The premium on legacy language skills decreases because AI can analyze COBOL even when humans can't. The premium on domain expertise increases because humans must verify that AI-generated specifications correctly capture business logic.[8] The ability to explain clearly, to verify carefully, to think systematically about complex systems matters more than memorizing syntax.

For government specifically, this shift is favorable. Government has deep domain expertise: the people who understand tax policy, benefits eligibility, regulatory compliance, procurement rules. What government has often lacked is the technical capacity to translate that expertise into working systems efficiently. When domain expertise becomes the critical input and AI handles the technical translation, government's existing strengths become more valuable. The people who understand how programs work become central to how technology gets built, not peripheral consultants brought in for requirements gathering and user acceptance testing.

The role of domain experts changes accordingly. Chapter 14 described how SpecOps puts domain expert verification at the center of the methodology rather than the periphery.[9] This isn't a minor process adjustment. It's a fundamental shift in who participates in technology work and how. Domain experts need time allocated for verification. Their input needs to be treated as authoritative, not advisory. The collaboration between technical staff and domain experts needs to become routine rather than exceptional.

Organizations that make this shift build capability that compounds. Each project that successfully integrates domain expertise builds relationships and practices that make the next project easier. Technical staff learn to write specifications that domain experts can engage with. Domain experts learn how to provide feedback that technical staff can act on. The organization develops a shared language and shared practices that span the traditional divide between "the business" and "IT."

Time is Running Out

The window for action is open, but it is narrowing. The workforce crisis makes knowledge capture urgent. The technology capability makes it possible. What we do in the next few years determines whether we break the cycle or repeat it.

Chapter 1 described a convergence: the workforce crisis reaching an acute phase while AI capabilities reached a threshold of usefulness for exactly the kind of knowledge extraction that legacy modernization requires.[10] This convergence creates a window. The domain experts who can verify specifications are still available, but not for much longer. The AI tools that can analyze legacy code at scale are now mature enough for production use.

Chapter 10's workforce data underscored the timeline: over 28 percent of federal workers are 55 or older, departures accelerated dramatically in 2025, and the average COBOL programmer is between 50 and 70 years old.[11] Every month of delay means more expertise lost. Every retirement takes institutional knowledge that could have been captured and preserved. The window is open now. It won't stay open indefinitely.

We simply do not have any more time to lose.

The alternative to breaking the cycle is continuing it. More aging systems. More crisis-driven modernizations. Billions more spent on replacements that become tomorrow's legacy problems. More knowledge lost when experts retire without their understanding being captured. More moments like the one in

New Jersey in April 2020, when critical systems fail under stress because they've become too fragile and too poorly understood to adapt.

The next crisis will find whatever systems exist at that moment. If those systems are still running on code no one fully understands, maintained by a shrinking pool of experts, documented by artifacts that long ago drifted from reality, the outcomes will be similar to what we've seen before. If those systems have been evolving continuously, guided by verified specifications that preserve institutional knowledge in durable form, the outcomes can be different. Not perfect, but different. Manageable rather than catastrophic.

This book has argued for a fundamental shift in how we think about government technology: from code as the valuable artifact to specification as the valuable artifact; from periodic crisis-driven replacement to continuous evolution; from knowledge locked in implementation details to knowledge preserved in forms that persist across technology generations.

This shift won't happen through a single dramatic transformation. There will be no announcement that government has adopted specification-driven development and the legacy crisis is solved. It will happen through thousands of decisions made by teams across federal, state, and local government: to invest in knowledge capture before it's too late, to verify specifications rather than just translate code, to treat systems as products rather than projects, to build for successors rather than just for current users.

It will happen quietly, incrementally, in the same way that DevOps transformed operations and GitOps transformed infrastructure management. One team at a time. One agency at a time. One successful modernization that preserves knowledge rather than just replacing code.

The legacy systems that governments struggle with today were, in their time, modern investments built by people who were solving real problems with the best tools available. The COBOL programs running unemployment insurance were cutting-edge in the 1960s and 70s. The mainframes processing Social Security benefits represented transformative capability. These systems succeeded so well that enormous institutional investments were built around them. Success became a constraint. Constraint became a crisis.

The systems we build today will follow the same arc. The languages we choose will fall out of fashion. The frameworks we adopt will be superseded. The cloud platforms we deploy to will evolve or be replaced. The people who build these systems will eventually move on. The only question is whether we leave behind code that future teams will struggle to understand, or specifications that preserve knowledge in forms that can guide whatever comes next.

Breaking the legacy cycle isn't about choosing better technology. Technology changes regardless of our choices. Breaking the cycle is about changing what we preserve. When we preserve specifications, we preserve understanding. When we preserve understanding, we enable evolution rather than crisis.

The next chapter looks outward: at what this approach means beyond government, at the broader implications for software development, at what needs to happen to build the community of practice that can make specification-driven development the norm rather than the exception.

But the core issue is already clear. We have the tools. We have the methodology. We have a narrowing window of opportunity. What we do with the opportunity we have is up to us.

Chapter 17: Beyond SpecOps - The Path Ahead

This book has focused on government, but the dynamics it describes play out everywhere legacy systems run. A hospital network operates patient records on systems built when fax machines were cutting-edge. An insurance company processes claims through mainframe applications that predate the people now maintaining them. A manufacturing firm runs its supply chain on software older than most of its workforce. Banks, utilities, transportation companies, healthcare providers: the same patterns appear across different sectors.

Decades-old systems. Retiring expertise. Knowledge locked in code no one fully understands. The specific technologies differ. The policy contexts differ. But the underlying challenge is the same: critical systems have become opaque, and the people who could explain them are leaving.

The approach developed throughout this book has applications beyond government. Its broader adoption could change how we think about software systems generally. This chapter explores those wider implications and then turns to what needs to happen to make them real.

Beyond Government

Healthcare, financial services, insurance, and utilities all operate systems built decades ago. These sectors share many characteristics with government: long system lifespans, complex regulatory requirements, high stakes for errors, and risk aversion that discourages change. A hospital running patient records on 1990s technology faces the same knowledge loss problem as a state agency running a benefits eligibility system on COBOL. The compliance requirements differ, but the dynamic is familiar.

Financial services institutions run enormous legacy portfolios. Core banking systems, payment processing, risk management: much of the financial infrastructure runs on mainframe applications built when those systems were state-of-the-art. The expertise to maintain these systems is aging out of the workforce. The knowledge is concentrated in a shrinking number of people.

Large-scale system replacements fail in the private sector too. SAP implementations run years over schedule and billions over budget. ERP migrations disrupt operations for months. CRM consolidations lose customer data. The private sector has more flexibility in some dimensions but faces the same fundamental challenges.

The difference is visibility. Government failures become headlines. Private sector failures often become write-offs and executive departures. But the underlying pattern is the same: treating code as the valuable artifact, allowing knowledge to concentrate in individuals, and attempting periodic big-bang replacements when systems become untenable.

The core challenge SpecOps addresses exists wherever legacy systems run: knowledge locked in aging code, expertise concentrated in a shrinking workforce, documentation that has drifted from reality. Part IV of this book explained why SpecOps fits government's specific context: the accountability requirements, the politics of modernization, the collaboration opportunities across agencies. But the underlying approach transfers beyond that context.

A healthcare organization could use the same methodology to modernize patient records systems, with clinical staff verifying that specifications correctly capture medical workflows. A bank could apply it to core banking modernization, with operations experts verifying business rules. An insurance company could use it for claims processing systems, with actuaries and underwriters verifying calculation logic.

The domain expertise changes. The regulatory context differs. The specific dynamics that Part IV examined for government would manifest differently in other sectors. But the fundamental methodology applies wherever the knowledge preservation problem exists: preserve knowledge in specifications that domain experts can verify, use AI to make comprehensive knowledge capture practical, and enable continuous evolution rather than periodic crisis.

Implications for Software Development

Specification-driven development represents more than a legacy modernization technique. It offers a different way of thinking about software that addresses problems created by decades of treating code as the primary artifact.

For most of software development's history, source code has been the authoritative record of system behavior. Documentation is secondary, often neglected, and commonly out of date. When you want to know what a system really does, you read the code.

This sometimes works when systems are small, developers stay with projects, and technologies remain stable. It breaks down over time. Systems grow complex. People leave. Technologies become obsolete. Code becomes the only record of decisions made years ago by people no longer available to explain them. The problems this book has examined in government legacy systems are the logical endpoint of code-as-artifact thinking applied over decades.

Treating specifications as the source of truth inverts this phenomenon. The specification describes what the system should do. The code implements the specification. When they diverge, the specification is authoritative. Changes flow through specifications first.

This approach has some intellectual heritage. Test-driven development established the discipline of describing expected behavior before writing implementation code.[1] Behavior-driven development extended this with Gherkin specifications: human-readable descriptions of system behavior that serve as both documentation and executable tests.[2] Formal methods, design-by-contract, and model-driven development have explored similar territory from different angles. SpecOps shares DNA with these approaches. The common thread is that describing system behavior precedes building implementation.

What's new is that AI makes maintaining comprehensive specifications practical in ways it wasn't before. The burden of keeping specifications synchronized with code has historically made specification-driven approaches impractical for most teams at scale. Writing and maintaining detailed specifications for large systems required effort that competed with actually building the systems. AI changes that calculation. Extraction, generation, and maintenance of specifications become tractable in ways they weren't when every word required human authorship.

The approach developed for legacy modernization could apply to new development as well. Start with specifications. Use AI to generate implementations. Verify that implementations match specifications. When requirements change, update specifications first.

The approach developed for legacy modernization applies to new development as well. Chapter 5 explored this territory: start with specifications, use AI to generate implementations, verify that implementations match specifications, update specifications first when requirements change. The same discipline that makes legacy systems comprehensible can make new systems maintainable from the start. Organizations that adopt specification-driven development for modernization may find themselves applying it more broadly because the practice proves its value.

Who Gets to Participate

Specification-driven development has the potential to open civic technology to people who haven't had a clear path to contributing. The approach values domain expertise over coding skills, creating opportunities for meaningful participation by people who understand policy, operations, and government processes even if they've never written code.

The civic technology movement has made enormous contributions to improving government services. Digital service teams at federal and state levels, civic tech organizations, and local civic tech communities have brought modern development practices to government technology. But participation has largely required technical skills. Hackathons need hackers. Digital service teams need developers. The work has been done primarily by people who can write code.

This has created a community that skews young, technical, and tends to be concentrated in major metropolitan areas where tech talent clusters. Civic tech hasn't intentionally excluded others. The work has simply required skills that not everyone has, and people whose expertise lies elsewhere haven't always seen where they might fit in.

Specification-driven development shifts what skills matter. When specifications are the valuable artifact, domain expertise becomes more valuable than coding ability. The people who can verify whether a specification correctly captures benefits eligibility rules aren't necessarily programmers. They're the people who understand benefits policy. The people who can confirm that a specification accurately describes tax calculations aren't necessarily developers. They're the people who understand tax law and agency procedures.

This could open civic technology to participants who might not have felt they could contribute before. A retired caseworker with decades of experience processing claims has knowledge that's essential for verification. A policy analyst who has never written a line of code can evaluate whether specifications match regulatory intent. An operations manager who knows how frontline staff actually use systems can identify gaps between documented and actual workflows.

Civic tech has skewed young partly because the technical skills it requires are concentrated among younger workers who grew up with technology. Specification-driven development creates opportunities for older participants whose value lies in institutional knowledge rather than coding skills.

This matters beyond diversity for its own sake. The people who haven't seen a clear path to contributing are often the people who understand government systems most deeply. They've spent careers learning how programs actually work, where the workarounds are, why systems behave the way they do. That knowledge is exactly what specification-driven development needs for verification. The expertise that has been walking out the door through retirements could instead be captured through participation in modernization efforts.

The civic tech community has been trending toward broader participation for years, with designers, researchers, and product managers joining what was once primarily a developer community. Specification-driven development can accelerate this trend and extend it further, to include participants whose expertise is in government operations rather than technology at all.

What Needs to Happen

Realizing the potential of specification-driven development requires action from multiple audiences: practitioners who do the work, leaders who create conditions for success, policymakers who shape the broader environment, and researchers who build the evidence base.

For practitioners, the message is simple: start now, start small. The methodology is described in this book. The tools now exist. Teams can begin using specification-driven approaches today without waiting for policy changes or organizational transformation. Chapter 13 provided practical guidance for getting started: selecting a pilot component, assembling a team, setting up tools. The first step is simply to begin.

Start with a manageable component where domain expertise is available. Work through the methodology. Learn what works in your context. Build evidence that the approach delivers value. Small successes create momentum for larger efforts.

Don't wait for perfect conditions. This book has been honest that challenges exist: domain experts may be scarce, legacy code may be incomprehensible, stakeholders may demand impossible timelines. Chapter 15 addressed strategies for imperfect conditions. Perfection isn't required. Progress is.

For leaders, the imperative is to invest in knowledge capture before it's too late. The workforce crisis is not hypothetical. It's happening right now. Every month that passes, more institutional knowledge walks out the door. Leaders who defer investment in knowledge capture may find they've waited too long.

This means funding knowledge preservation as an explicit priority, not an afterthought. It means allocating domain experts time for verification work. It means creating incentives for retiring staff to participate in knowledge capture. It means treating specifications as organizational assets that deserve investment.

Leaders also need to protect teams from pressure to skip verification or rush through the methodology. The shortcuts that seem expedient in the moment create problems that can grow larger over time. Chapter 15 discussed how to have conversations about tradeoffs. Leaders need to ensure those conversations happen rather than allowing schedule pressure to override quality.

For policymakers, the opportunity is to address structural barriers. Budget structures that treat systems as projects rather than products create pressure against continuous investment. Procurement rules that favor large contracts over modular approaches make incremental modernization difficult. Workforce policies that make it hard to retain and recruit talent with needed skills constrain what's possible.

Policy changes that would help include budget flexibility that allows continuous investment in system evolution, procurement approaches that enable modular contracting and incremental delivery, and workforce strategies that value domain expertise alongside technical skills.

The SHARE IT Act and similar initiatives recognize the importance of sharing code across government. Expanding these frameworks to encompass instruction sets and specifications would accelerate the collaboration Chapter 12 described.

For the research community, the need is to study what works. The approach this book describes is new enough that evidence is still developing. Rigorous research on specification-driven development would help build the evidence base that leaders and policymakers need.

Research questions abound. How do modernization outcomes differ when specifications are treated as the source of truth? What factors predict success or failure in specification-driven modernization? How does domain expert involvement affect specification quality? What makes instruction sets transferable across organizations? How do AI capabilities need to evolve to better support the methodology? Academic institutions, think tanks, and government research offices all have roles to play in answering these questions.

The SpecOps Community

Individual organizations adopting specification-driven development is valuable. A community of practice sharing knowledge and building on each other's work would be transformative.

Chapter 12 explained why instruction set sharing can work where code sharing has not yet: instruction sets operate at a layer where commonality exists regardless of jurisdictional policy differences. COBOL is COBOL everywhere. That chapter envisioned a shared repository of instruction sets organized by category, with agencies contributing what they've learned and benefiting from accumulated knowledge.

That vision remains to be realized. The infrastructure exists. What's needed is a community that actively contributes and maintains shared resources.

A mature community of practice around specification-driven development would include practitioners sharing instruction sets refined through actual use, case studies documenting successes and failures, training resources that help new teams get started, and forums for troubleshooting challenges. It would span government levels and eventually extend to other sectors facing similar challenges.

The open source model provides a template. Projects succeed when communities form around them, contributing improvements and extending capabilities. The same dynamic could apply to instruction sets and system specification patterns. Early contributors create value for everyone who follows after them.

The book has provided methodology, principles, and practical guidance. What it can't provide is the community that makes isolated efforts into a movement. That requires people who do the work, share what they learn, and build on each other's contributions.

The civic technology community has demonstrated this is possible. Local civic tech communities, govtech conferences, open source government projects: these exist because people decided to participate. Specification-driven development needs the same energy applied to a new challenge.

This book has made the argument that the legacy modernization crisis is real but addressable, that preserving knowledge matters more than translating code, that AI creates new possibilities for capturing institutional understanding before it disappears, and that specifications can serve as durable artifacts that outlast any particular technology implementation.

Since we've now reached the end, what matters now is what readers do with this information.

For some readers, that means starting a pilot project: selecting a component, assembling a team, and working through the methodology to see what it produces in their context. For others, it means advocating for investment in knowledge capture before the window closes further. For others still, it means contributing to the community of practice: sharing instruction sets, documenting what works, building the collective resources that make each subsequent effort easier.

The window is still open, but the opening is getting narrower.

Every specification verified is knowledge preserved. Every instruction set shared is a contribution to collective capability. Every team that demonstrates success creates evidence that builds momentum for broader adoption. The work happens one component at a time, one team at a time, one organization at a time.

Whether the approach takes hold depends on whether enough people do the work to make it matter. That's not something this book can determine.

Only its readers can.

Appendix A: SpecOps Quick Reference

This appendix provides a condensed reference for the SpecOps methodology, including the six phases, core principles, guidance for common decisions, and definitions of key terms used throughout this book.

The Six Phases

Phase 1: Discovery and Assessment. Understand the scope of the legacy system, identify stakeholders, assess available resources, and create a prioritized plan for specification development. Key activities include system inventory, stakeholder identification, knowledge assessment, component prioritization, tool selection, and pilot selection.

Phase 2: Specification Generation. Use AI agents with appropriate instruction sets to analyze legacy code and generate initial specifications describing system behavior. This phase produces draft specifications that capture what the legacy system does, including business logic, data flows, and edge cases.

Phase 3: Specification Verification. Domain experts review and validate specifications for correctness. This is the critical checkpoint where people who understand the business rules confirm that the AI-generated specifications accurately describe system behavior. Specifications are revised until domain experts approve them.

Phase 4: Modern Implementation. Generate new code from verified specifications. AI coding agents, guided by implementation instruction sets, produce modern code that implements the behavior described in the specifications. Engineers review and refine the generated code.

248

Phase 5: Testing and Validation. Verify that the modern implementation correctly implements the specification. Testing compares modern system behavior against specification requirements, not just against legacy system outputs. This phase includes unit testing, integration testing, and user acceptance testing.

Phase 6: Deployment and Knowledge Transfer. Move the modern implementation to production and ensure the organization can maintain it. This phase includes data migration, production deployment, staff training, specification finalization, and legacy system decommissioning.

Core Principles Checklist

When evaluating whether a modernization effort follows SpecOps principles, consider these questions:

Specification as source of truth. Is the specification repository treated as the authoritative description of system behavior? Do decisions reference the specification rather than the legacy code or new implementation?

Knowledge preservation precedes translation. Has the team invested in creating and verifying specifications before generating modern code? Would the specifications remain valuable even if modernization stalled?

Domain experts as arbiters. Are policy specialists and business analysts reviewing specifications for correctness? Is their input treated as authoritative, not advisory?

AI assists, humans verify. Is AI being used for code analysis and generation while humans verify correctness? Are there checkpoints where domain experts confirm AI output?

Changes flow through specifications. Do system modifications begin with specification updates? Are specification changes reviewed before implementation?

Specifications are accessible. Can non-technical stakeholders read and understand the specifications? Are specifications written in plain language with technical details clearly separated?

Decision Guidance

When to use SpecOps versus direct translation. Use SpecOps when institutional knowledge is at risk, when domain experts are available to verify specifications, when long-term knowledge preservation matters, and when the system will be maintained for decades. Consider direct translation only for straightforward code conversion where business logic is simple and well-documented, and where speed matters more than verification.

When to start specification work. Begin specification work as soon as domain experts are available and legacy code is accessible. Do not wait for perfect tooling or complete documentation.

When specifications are ready for verification. Specifications are ready when they describe system behavior at a level domain experts can understand and verify, when technical terms are defined, when edge cases are documented, and when areas of uncertainty are flagged.

When modern implementation is ready for deployment. Modern implementation is ready when it passes all tests derived from specifications, when domain experts have validated behavior, when staff have been trained, and when rollback procedures are in place.

Glossary of Key Terms

AI coding agent. Software that uses artificial intelligence to assist with code analysis, specification generation, and code generation. Examples include Claude, GitHub Copilot, and Cursor.

Compilation (in SpecOps context). The process of extracting knowledge from legacy code and capturing it in specifications before generating modern code. Contrasts with transpilation.

Domain expert. A person who understands the business rules, policies, and intended behavior of a system. In government contexts, often policy specialists, program administrators, or business analysts.

Instruction set. A collection of guidance and context provided to AI coding agents to help them understand specific legacy platforms, domains, or specification formats. Also called "skills" or "custom instructions."

Legacy system. Software that remains in use despite being built on older technology, often because it performs critical functions that can't easily be replaced.

Specification. A detailed description of what a system does, written in a form that domain experts can verify and that guides modern implementation. In SpecOps, specifications are the authoritative source of truth for system behavior.

Specification repository. A version-controlled repository where specifications are stored, versioned, and managed. Functions as the single source of truth for system behavior.

SpecOps. Specification-Driven Operations. A methodology for modernizing legacy systems using AI to generate comprehensive, human-verified specifications that guide modern implementation.

Strangler Fig pattern. An approach to legacy modernization where new components are built alongside the legacy system, gradually replacing it piece by piece until the legacy system can be retired. Named after vines that gradually replace their host trees.

Transpilation. Direct translation of code from one language to another without extracting and verifying the underlying knowledge. Contrasts with compilation in the SpecOps approach.

Verification. The process by which domain experts review specifications or implementation to confirm correctness. In SpecOps, verification happens at the specification level where domain experts can meaningfully participate.

Appendix B: Starting Templates

This appendix provides templates to help teams begin a SpecOps modernization initiative. These templates are starting points; adapt them to fit your organization's context and needs.

Component Prioritization Matrix

Use this matrix to evaluate and rank legacy system components for modernization. Score each component on a scale of 1 (low) to 5 (high) for each criterion, then calculate the total.

Criterion	[Component A]	[Component B]
Knowledge Risk		
Business Value		
Technical Debt		
Complexity		
Domain Expert Availability		
Clear Boundaries		
Total Score		

Scoring guidance:

Knowledge Risk: How endangered is the knowledge needed to maintain this component? Score 5 if the only expert has retired or will soon; score 1 if knowledge is well-documented and widely held.

Business Value: How much value would modernizing this component deliver? Score 5 for core business-critical functions; score 1 for rarely-used utilities.

Technical Debt: How much maintenance burden does this component create? Score 5 for frequent incidents and workarounds; score 1 for stable, low-maintenance code.

Complexity: How complex is this component? Score inversely for pilot selection: score 5 for moderate complexity suitable for pilots; score 1 for extremely complex or trivial components.

Domain Expert Availability: Are experts available to verify specifications? Score 5 if multiple experts are available and committed; score 1 if no experts exist.

Clear Boundaries: How well-defined are this component's integration points? Score 5 for components with clean interfaces; score 1 for deeply entangled components.

Specification Template

Use this structure for component specifications. Not all sections will apply to every component; mark inapplicable sections as "Not applicable" rather than omitting them.

```
# [Component Name] Specification

Version: [X.X]

Status: [Draft | Under Review | Verified | Approved]

Last Updated: [Date]

## 1. Overview

 - Component name and identifier

 - Purpose and responsibilities

 - Position in overall system architecture

 - Key stakeholders
```

2. Business Context

- What business need this component serves

- Who uses it and how

- What business processes it supports

- Relevant policies or regulations

3. Inputs

- Data received (structure, source, frequency)

- Events that trigger processing

- Dependencies on other components

- External system inputs

4. Business Logic

- Rules and decision points

- Calculations and formulas

- Validation requirements

- Exception handling

5. Outputs

- Data produced (structure, destination, frequency)

- Events triggered

- Side effects and state changes

- Reports or notifications generated

6. Data Models

- Internal data structures

- Database schemas

- File layouts

- Data relationships and constraints

7. Integration Points

- Other components this depends on

- Components that depend on this

- External system integrations

- APIs or interfaces exposed

8. Edge Cases and Exceptions

- Unusual scenarios and how handled

- Known limitations

- Workarounds for specific situations

9. Uncertainties and Questions

- Areas where specification may be incomplete

- Ambiguities requiring domain expert clarification

- Assumptions that need validation

10. Verification Status

- Who has reviewed this specification

- What sections have been verified

- Outstanding issues or questions

- Approval status

```
## 11. References

- Related specifications

- Legacy code locations

- Policy documents
```

Instruction Set Template

Use this structure when creating instruction sets (skills) for AI coding agents.

```
# [Instruction Set Name]

## Purpose

[One paragraph describing what this instruction set helps AI
agents accomplish]

## When to Use

[Describe the scenarios where this instruction set applies]

## Key Concepts

[List and briefly explain the key concepts the AI needs to
understand]

## Patterns to Recognize

[Describe code patterns, structures, or indicators the AI
should look for]

## Extraction/Generation Guidelines

[Step-by-step guidance for how the AI should process what it
finds]

## Output Format
```

```
[Describe how results should be structured and formatted]

## Common Pitfalls

[List mistakes the AI commonly makes and how to avoid them]

## Examples

[Provide concrete examples of input and expected output]

## Quality Checks

[List criteria for evaluating whether output is acceptable]
```

Verification Checklist

Use this checklist when domain experts review specifications for correctness.

Before Review

- [ ] Specification is complete (all applicable sections filled in)
- [ ] Specification has been technically reviewed by engineers
- [ ] Supporting materials are available (policy documents, test cases, examples)
- [ ] Reviewer has adequate time allocated for thorough review

During Review

- [ ] Business rules are correctly stated
- [ ] Calculations and formulas are accurate
- [ ] Edge cases are properly identified and handled
- [ ] Terminology matches how domain experts describe the system
- [ ] Policy implementation is accurate
- [ ] Nothing is missing that should be included

- [ ] Nothing is included that contradicts actual system behavior

Issues to Flag

- [ ] Incorrect business logic
- [ ] Missing edge cases
- [ ] Policy misinterpretations
- [ ] Terminology errors
- [ ] Incomplete coverage
- [ ] Contradictions between sections
- [ ] Areas requiring additional clarification

After Review

- [ ] All issues documented with specific locations and descriptions
- [ ] Severity of each issue indicated (critical, major, minor)
- [ ] Suggestions for correction provided where possible
- [ ] Overall assessment recorded (approve, approve with changes, requires revision)
- [ ] Reviewer name and date recorded

Appendix C: Tool Landscape

This appendix provides an overview of tools relevant to SpecOps implementation. The AI coding tool market evolves rapidly; specific capabilities and recommendations may change. Use this as a starting point for evaluation rather than a definitive guide.

AI Coding Tools

Several AI coding assistants can support SpecOps work. Each has different strengths; many teams use multiple tools for different phases of the work.

Claude (Anthropic). Strong at specification generation, reasoning about complex business logic, and following detailed instructions. Large context window allows processing substantial code and specifications in a single session. Available through web interface, API, and integrated into some development environments.

GitHub Copilot (GitHub/Microsoft). Tightly integrated with common development environments. Supports multiple underlying models including Claude, Gemini, and GPT-4. Widely available in enterprise settings. Strong at code completion and generation.

Cursor (Anysphere). AI-native integrated development environment. Designed around AI assistance from the ground up rather than as an add-on. Strong at multi-file editing and codebase-wide changes.

Aider (Open source). Command-line tool for AI pair programming. Works with multiple AI models. Useful for teams preferring terminal-based workflows.

Continue.dev (Open source). Extensible AI code assistant that integrates with VS Code and JetBrains IDEs. Supports multiple model providers.

When selecting tools, consider context window size (can the tool handle large specifications and codebases?), instruction following (how well does it adhere to custom instruction sets?), legacy language support (does it understand COBOL, RPG, or other legacy languages?), integration with existing workflows, and cost.

Recommended Toolchain Configurations

Minimum Viable Configuration. Suitable for pilot projects and teams just getting started.

Category	Tool	Purpose
Version Control	GitHub or GitLab	Store specifications, enable collaboration
AI Coding Agent	Any tool listed above	Analyze legacy code, generate specifications
Editor	VS Code or similar	Author and edit specifications
Review Workflow	GitHub Pull Requests	Review and approve specification changes

This configuration costs little beyond existing subscriptions and can be operational within a day. It provides version control, AI assistance, and a review process.

Mature Configuration. Suitable for teams with established SpecOps practice.

Category	Tool	Purpose
Version Control	GitHub or GitLab	Store specifications with branch protection
AI Coding Agent	Multiple tools for different tasks	Specialized analysis and generation
Editor	AI-native IDE (Cursor or similar)	Integrated AI assistance
Review Workflow	Pull requests with required reviewers	Formal verification workflow
Automated Validation	Custom linters and scripts	Check specification completeness
Testing Infrastructure	pytest, JUnit, or similar	Verify implementations
CI/CD	GitHub Actions or similar	Automate validation and regeneration
Documentation	MkDocs, GitBook, or similar	Publish specifications

This configuration requires more setup but provides automation, formal workflows, and published documentation.

Open Source Resources

SpecOps Methodology Documentation. The SpecOps methodology documentation is available as an open source repository. It includes the manifesto, methodology guide, team structure guidance, instruction set examples, and FAQ.

spec-kit (GitHub). Specification-driven development tooling from GitHub, providing templates and validation tools for specification workflows.

Aider. Open source command-line AI pair programming tool that works with multiple AI model providers.

Continue.dev. Open source AI code assistant with extensibility for custom workflows.

Community Resources

Government Technology Communities. Several communities focus on government technology modernization and share experiences, including challenges and solutions for legacy system work.

AI Coding Tool Communities. Most AI coding tools have active user communities where practitioners share instruction sets, prompts, and approaches. These communities often share techniques applicable to legacy code analysis.

Legacy Language Communities. Communities focused on COBOL, RPG, and other legacy languages continue to exist and can provide insight into language-specific patterns and challenges.

Selection Considerations

When building your toolchain, remember that the best tool is the one your team will actually use. Start with tools your team already knows when possible. Invest in training for tools that offer significant advantages. Expect to iterate on your toolchain as you learn what works.

Don't wait for perfect tooling. The minimum viable configuration can support meaningful SpecOps work. Add sophistication as your practice matures and you understand what additional automation would help most.

Appendix D: Further Reading

This appendix provides selected sources for readers who wish to explore topics covered in this book more deeply. The sources are organized by theme rather than by chapter, allowing readers to pursue the subjects most relevant to their interests and needs.

Legacy Code and Modernization

Marianne Bellotti, *Kill It with Fire: Manage Aging Computer Systems (and Future Proof Modern Ones)* (San Francisco: No Starch Press, 2021). Drawing on her experience at the U.S. Digital Service, Bellotti provides a practitioner's perspective on legacy system analysis, the Strangler Fig pattern, and the organizational dynamics that shape modernization outcomes. Bellotti's central insight—that legacy systems deserve respect before replacement—aligns closely with the SpecOps emphasis on understanding before transformation.

Michael Feathers, *Working Effectively with Legacy Code* (Upper Saddle River, NJ: Prentice Hall, 2004). The definitive practitioner's guide to working with inherited codebases. Feathers' concepts of characterization tests and seams provide the intellectual foundation for incremental legacy modernization approaches including SpecOps. His definition of legacy code as "code without tests" reframes the problem in terms of verifiability rather than age.

Frederick P. Brooks Jr., *The Mythical Man-Month: Essays on Software Engineering* (Reading, MA: Addison-Wesley, 1975; Anniversary Edition, 1995). Brooks' classic collection of essays on software engineering remains essential reading. His discussion of the "second-system effect"—the tendency to overload replacement systems with deferred features—is particularly relevant to understanding why government modernization projects often fail.

Martin Fowler, "Strangler Fig Application," martinfowler.com, August 22, 2024.
https://martinfowler.com/bliki/StranglerFigApplication.html
Fowler's influential blog post describes the pattern of incrementally replacing legacy systems by building new functionality alongside old, gradually routing more traffic to the new system until the legacy can be retired. This pattern is central to how SpecOps approaches large-scale modernization.

Government Technology and Digital Services

U.S. Government Accountability Office, "Information Technology: Agencies Need to Plan for Modernizing Critical Decades-Old Legacy Systems," GAO-25-107795 (July 2025). https://www.gao.gov/products/gao-25-107795
The most comprehensive recent assessment of federal legacy systems, profiling eleven critical systems ranging from 23 to 60 years old with combined annual operating costs of $754 million. Essential reading for understanding the scope and urgency of the government legacy modernization challenge.

Office of the Auditor General of Canada, "Building and Implementing the Phoenix Pay System" (May 2018).
https://www.oag-bvg.gc.ca/internet/English/parl_oag_201805_01_e_43033.html A sobering case study in how legacy modernization can go catastrophically wrong. The Phoenix payroll system caused pay problems for nearly 80 percent of Canada's 290,000 federal public servants and continues to generate issues years after launch. The report illuminates failures in governance, testing, and domain expert involvement that SpecOps is designed to address.

U.S. Digital Service, *Digital Services Playbook.*
https://playbook.cio.gov/ Developed by the U.S. Digital Service, this playbook provides thirteen principles for building effective digital services. Plays 4 (agile and iterative practices) and 7 (experienced teams with subject matter expertise) are particularly relevant to SpecOps implementation.

18F, *De-risking Government Technology Guide.*
https://www.gsa.gov/blog/2024/09/11/a-revised-and-expanded-guide-for-derisking-government-technology-projects A practical guide for state agencies and their federal partners on how to structure and oversee custom technology projects to reduce risk. Covers budgeting, procurement, and project oversight from a practitioner perspective.

AI and Software Development

Gene Kim and Steve Yegge, *Vibe Coding: Building Production-Grade Software With GenAI, Chat, Agents, and Beyond* (Portland, OR: IT Revolution, 2025). A comprehensive exploration of how generative AI is transforming software development practices. The authors examine both the opportunities and risks of AI-assisted coding, with practical guidance for building organizational culture around these new capabilities.

Sean Grove, "The New Code," AI Engineer conference talk (2025).
https://www.youtube.com/watch?v=8rABwKRsec4 Grove, who works on alignment research at OpenAI, articulates the concept of specifications as source artifacts that should be preserved and version-controlled, with code as derivative output that can be regenerated. This talk provides key conceptual foundations for the SpecOps approach.

MITRE, "Legacy IT Modernization with AI" (June 2025).
https://www.mitre.org/news-insights/publication/legacy-it-mod ernization-ai MITRE's research found that large language models can reliably generate intermediate representations at scale from legacy code, but automated metrics do not match human expert assessments of quality. This empirically validates the SpecOps principle that domain experts must verify AI-generated specifications.

GitHub, "How GitHub Copilot and AI agents are saving legacy systems," *The GitHub Blog* (October 2025).
https://github.blog/news-insights/research/how-github-copilot-and-ai-agents-are-saving-legacy-systems/ An industry perspective on how AI coding assistants are being applied to legacy modernization challenges, with case studies and practical insights.

Microsoft Azure Blog, "How We Use AI Agents for COBOL Migration and Mainframe Modernization" (July 2025).
https://azure.microsoft.com/en-us/blog/how-we-use-ai-agents-f or-cobol-migration-and-mainframe-modernization/ A candid assessment from practitioners describing their COBOL migration experiments as producing "a good mix of educated guesses (from us and the model) and hallucinational gibberish"—an honest acknowledgment of both capabilities and limitations.

GitOps and Infrastructure as Code

Alexis Richardson, "Operations by Pull Request," Weaveworks blog (March 2017).
https://web.archive.org/web/20230608145857/https://www.we ave.works/blog/gitops-operations-by-pull-request The original post describing the operational practices that became known as GitOps. Richardson's insight—that version-controlled declarative

descriptions should be the source of truth for infrastructure—directly inspired the SpecOps approach to specifications.

Weaveworks, "What Is GitOps Really?" (May 2019).
https://web.archive.org/web/20230531033515/https://www.weave.works/blog/what-is-gitops-really The four principles of GitOps (declarative configuration, version control, automated deployment, and continuous reconciliation) have been adopted by the broader GitOps community and inform the SpecOps methodology.

The Government Workforce Challenge

Partnership for Public Service, "A Profile of the 2023 Federal Workforce."
https://ourpublicservice.org/fed-figures/workforce/ Analysis of federal workforce demographics using OPM FedScope data. Documents the age distribution that makes legacy knowledge preservation urgent: 28 percent of full-time permanent federal workers are age 55 or above, compared to about 23 percent in the private sector.

USAFacts, "Six charts on the age of federal workers" (April 2025).
https://usafacts.org/articles/six-charts-on-the-age-of-federal-workers/ Visualization of federal workforce age trends from 2005 to 2024, showing the persistent concentration of federal employees in older age brackets and the underrepresentation of workers under 30.

U.S. Census Bureau, "By 2030, All Baby Boomers Will Be Age 65 or Older" (December 2019).
https://www.census.gov/library/stories/2019/12/by-2030-all-baby-boomers-will-be-age-65-or-older.html Context for the broader

demographic transition affecting the government workforce, with approximately 10,000 Baby Boomers reaching retirement age each day.

The SpecOps Methodology

SpecOps Method Repository.
https://github.com/spec-ops-method/spec-ops The open-source repository containing the complete SpecOps methodology documentation, including the Manifesto, Methodology phases, Team Structure guidance, Instruction Sets framework, FAQ, and Comparison Framework. This is the authoritative source for methodology details beyond what this book covers.

Mark Headd, "SpecOps: A New Approach to Legacy System Modernization," Civic Innovations (November 2025).
https://civic.io/2025/11/24/specops-a-new-approach-to-legacy-system-modernization/ An introduction to the SpecOps approach and its origins, written for a general audience interested in government technology.

Mark Headd, "What Does a Good Spec File Look Like?" Civic Innovations (December 2025).
https://civic.io/2025/12/02/what-does-a-good-spec-file-look-like/ A practical exploration of specification file structure and content, with examples from the IRS Direct File demonstration.

Case Studies and Examples

California State Auditor, "Department of Motor Vehicles: Its Automation Project Has Suffered Extensive Delays and Cost Overruns," Report 94-107 (April 1994).

https://www.bsa.ca.gov/reports/summary/94107
Documentation of California's first DMV modernization attempt (1987-1994), which spent $50 million over seven years before cancellation.

California Legislative Analyst's Office, "DMV: Withdraw Funding for Database Project" (2013).
https://lao.ca.gov/Recommendations/Details/743
Documentation of California's second DMV modernization attempt (2006-2013), which spent $135 million over seven years before cancellation. Together with the 1994 report, these documents illustrate the pattern of modernization failure that SpecOps aims to break.

Ad Hoc, "Using generative AI to transform policy into code" (September 2024).
https://adhoc.team/2024/09/25/using-generative-ai-to-transform-policy-into-code/ Description of the Policy2Code Prototyping Challenge approach to using large language models to generate intermediate representations between policy language and software code—an approach conceptually similar to SpecOps specification generation.

IRS Direct File Repository.
https://github.com/IRS-Public/direct-file The public repository for the IRS Direct File project, which provides source code that can be used to demonstrate SpecOps specification generation. A demonstration repository with sample skills files and generated specifications is available at https://github.com/spec-ops-method/spec-ops-demo.

Historical Context

Melvin Conway, "How Do Committees Invent?" *Datamation* **(April 1968).**

http://www.melconway.com/Home/Committees_Paper.html
The original statement of what became known as Conway's Law:
"Organizations which design systems are constrained to produce
designs which are copies of the communication structures of
these organizations." This insight helps explain why legacy
systems often reflect the organizational structures that created
them.

**Clay Shirky, "Getting to the Bottom of HealthCare.gov's
Flop," *The New York Times* (October 2013).**
https://www.nytimes.com/2013/10/25/opinion/getting-to-the-b
ottom-of-healthcaregovs-flop.html Analysis of the Healthcare.gov
launch failure that helped catalyze the creation of the U.S. Digital
Service and 18F, and shaped the modern government technology
reform movement.

Notes

Chapter 1

1. U.S. Government Accountability Office, "Information Technology: Agencies Need to Plan for Modernizing Critical Decades-Old Legacy Systems," GAO-25-107795 (July 2025), 45. COBOL (Common Business Oriented Language) was introduced in 1959 and became the first widely used, high-level programming language for business applications.

2. Thad Rueter, "Post-COVID, Unemployment Systems Take Stock and Go Modern," *Government Technology*, March 2023. https://www.govtech.com/computing/post-covid-unemployment-systems-take-stock-and-go-modern California's Employment Development Department faced billions in fraud perpetrated by individuals exploiting relaxed eligibility rules used to speed economic relief during the pandemic.

3. Ibid. Kansas Governor Laura Kelly stated in a letter to the U.S. Department of Labor: "Despite our unemployment rate returning to historically low levels, Kansas is still using antiquated equipment to work through pandemic-related claims."

4. GAO-25-107795, 1.

5. Ibid., 4. For fiscal year 2025, about $83 billion (79 percent) of the $105 billion in planned total IT spending for the 24 Chief Financial Officers Act agencies was allocated to operations and maintenance.

6. Ibid., 7. The GAO describes consequences of not updating legacy systems including cybersecurity risks, unmet mission needs, shortage of staff with specialized skills, and increased costs.

7. GAO-25-107795, 19-20.

8. Ibid., 18.

9. Ibid., 20. Both Treasury Department systems run on COBOL and Assembly Language Code, with the GAO noting that relying

on these languages carries risks such as rising procurement and operating costs and fewer individuals with proper skill sets.

10. Ibid. The Environmental Protection Agency's system contains obsolete hardware that is not supported by manufacturers and has known cybersecurity vulnerabilities that can't be remediated without modernization.

11. Ibid., 53. The Indian Health Service's system was originally implemented in 1969 and is based on the MUMPS programming language, with its last standard form approved 25 years ago, making it considered both legacy and obsolete.

12. Social Security Administration, Office of the Inspector General, "Modernizing Social Security's Information Technology Infrastructure," Congressional Testimony (July 14, 2016). https://www.ssa.gov/oig/congressional_testimony/Testimony-COBOL-IT-Infrastructure-071416.pdf "SSA maintains more than 60 million lines of COBOL today, along with millions more lines of other legacy programming languages."

13. "Individual Master File," Wikipedia, https://en.wikipedia.org/wiki/Individual_Master_File (accessed December 2025); Jack Hagel, "The IRS System Processing Your Taxes is Almost 60 Years Old," Nextgov/FCW, March 2018. https://www.nextgov.com/it-modernization/2018/03/irs-system-processing-your-taxes-almost-60-years-old/146746/ The IMF is written in Assembly Language Code and COBOL, with an architecture and design dating back to the 1960s.

14. TechChannel, "Special Report: COBOL Survey Results Prove Pervasiveness, Value, and a Bright Future," https://techchannel.com/cobol/special-report-cobol-survey-results-prove-pervasiveness-value-and-a-bright-future/ (accessed December 2025).

15. Tom Taulli, "COBOL Language: Call It A Comeback," *Forbes*, July 13, 2020, https://www.forbes.com/sites/tomtaulli/2020/07/13/cobol-language-call-it-a-comeback/

16. "Old Coding Language Entrenched in 100+ Government Systems," *ZoomInsights*, Medium, https://medium.com/zoominsights/old-coding-language-entrenc

hed-in-100-government-systems-1124e4bb9f36 (accessed
December 2025).

17. "Aging Workforce Brings COBOL Crisis," *SIGNAL Magazine*,
AFCEA,
https://www.afcea.org/signal-media/cyber-edge/aging-workforc
e-brings-cobol-crisis (accessed December 2025).

18. U.S. Census Bureau, "By 2030, All Baby Boomers Will Be Age
65 or Older," December 10, 2019,
https://www.census.gov/library/stories/2019/12/by-2030-all-ba
by-boomers-will-be-age-65-or-older.html

19. Partnership for Public Service, "A Profile of the 2023 Federal
Workforce," analysis of OPM FedScope data.
https://ourpublicservice.org/fed-figures/workforce/ See also
USAFacts, "Six charts on the age of federal workers," April 2025,
analyzing OPM data from September 2024.
https://usafacts.org/articles/six-charts-on-the-age-of-federal-wo
rkers/

20. "COBOL: a Demographic Disaster in the Making," *Open
Mainframe Project*, September 17, 2019.
https://lists.openmainframeproject.org/g/wg-cobol/attachment/
113/1/systemsjournalcobolsept172020.pdf

21. Ibid.

22. Andrew Starrs, testimony before House Ways and Means
Committee, "Social Security Administration Information
Technology Modernization" (March 7, 2018),
https://waysandmeans.house.gov/wp-content/uploads/2018/04
/20180307SS-Transcript.pdf

23. COBOL Cowboys, https://cobolcowboys.com/ (accessed
December 2025). COBOL Cowboys is a consulting firm that
maintains networks of veteran COBOL programmers and
connects agencies with retired experts for emergency system
maintenance.

24. Rueter, "Post-COVID, Unemployment Systems." Multiple
states including New Jersey brought back retired programmers
during the pandemic unemployment surge.

25. Ed Airey, "The Government's COBOL Conundrum," *FedTech Magazine*, June 2014. https://fedtechmagazine.com/article/2014/06/governments-cobol-conundrum According to the Office of Personnel Management's Strategic Information Technology Plan, OPM anticipates costs will increase 10 to 15 percent annually "as personnel with the necessary coding expertise retire and can't be easily replaced."

26. Clay Shirky, "Getting to the Bottom of HealthCare.gov's Flop," *The New York Times*, October 24, 2013. https://www.nytimes.com/2013/10/25/opinion/getting-to-the-bottom-of-healthcaregovs-flop.html

27. Ibid.

28. GAO-25-107795, 8. "As a result of the difficulties in acquiring, developing, and managing IT investments the federal government has experienced, we identified 'Improving the Management of IT Acquisitions and Operations' as a high-risk area in February 2015."

29. Office of the Auditor General of Canada, "Building and Implementing the Phoenix Pay System" (May 2018). https://www.oag-bvg.gc.ca/internet/English/parl_oag_201805_01_e_43033.html By July 2018, Phoenix had caused pay problems for close to 80 percent of the federal government's 290,000 public servants through underpayments, overpayments, and non-payments.

30. PSAC (Public Service Alliance of Canada), "Government downplays Phoenix pay system failures while workers continue to suffer nine years later," February 2025. https://psacunion.ca/government-downplays-phoenix-pay-system-failures-while As of January 2025, costs had exceeded $5 billion with 372,000 outstanding cases remaining.

31. Office of the Auditor General of Canada, "Building and Implementing the Phoenix Pay System" (May 2018). https://www.oag-bvg.gc.ca/internet/English/parl_oag_201805_01_e_43033.html The report found that Phoenix executives "decided to defer or remove more than 100 important pay processing functions" when costs exceeded budget, and documents the complexity of federal pay rules requiring more

than 80,000 rules and 200 custom programs added to PeopleSoft software.

32. Office of the Auditor General of Canada, "Building and Implementing the Phoenix Pay System" (May 2018). https://www.oag-bvg.gc.ca/internet/English/parl_oag_201805_01_e_43033.html Auditor General Michael Ferguson's characterization of the Phoenix project.

33. GAO-25-107795, 9. "The first three efforts reflect varying approaches that the department had taken since 2001 to achieve a modernized EHR system. However, these approaches were abandoned due to concerns about project planning, high costs, and length of time to deliver capabilities."

34. Ibid. "In April 2023, after deploying the new system to five of its medical centers, VA paused deployments due to feedback from veterans and clinicians that the new system was not meeting expectations."

35. Ibid. "We noted that although user satisfaction had improved over the last 3 years, users at the five initial sites continued to be generally dissatisfied with the new system."

36. U.S. Government Accountability Office, "VA Electronic Health Record Modernization: Significant Risks Need to Be Addressed to Achieve Program Goals," GAO-25-106874 (December 2024), https://www.gao.gov/products/gao-25-106874

37. California State Auditor, "Department of Motor Vehicles: Its Automation Project Has Suffered Extensive Delays and Cost Overruns," Report 94-107 (1994), https://information.auditor.ca.gov/reports/summary/94107

38. Skip Descant, "Is There Hope for Modernizing State DMVs?" *Government Technology*, March 2023. https://www.govtech.com/computing/is-there-hope-for-modernizing-state-dmvs Oregon completed a modernization that "retired nearly 100 old legacy systems."

39. Ibid. Texas requested $6.75 million from the state Legislature to develop documentation, identify system requirements, and define the DMV modernization project.

40. Jule Pattison-Gordon, "Where Are States on the Path to Upgrading ERP Systems?" *Government Technology*, March 2023. https://www.govtech.com/finance/where-are-states-on-the-path-to-upgrading-erp-systems Missouri's main system is more than 20 years old and "staff that are able to support it are getting ready to retire."

41. GAO-25-107795, 26. GAO has made numerous recommendations to address IT modernization challenges, though many remain unimplemented.

42. U.S. Government Accountability Office, "GAO Calls for Urgent Action to Address IT Acquisition and Management Challenges," Press Release, February 2015. https://www.gao.gov/press-release/gao-calls-urgent-action-address-it-acquisition-and-management-challenges

43. Treasury Inspector General for Tax Administration, "The IRS Must Develop and Implement a Comprehensive Data Governance Framework to Support Future Modernization Efforts," Reference Number 2025-20-050 (September 2025), https://www.tigta.gov/sites/default/files/reports/2025-09/2025208050fr.pdf

Chapter 2

1. This design goal—making technical artifacts readable by domain experts, not just programmers—anticipates a theme that will recur throughout this book. The SpecOps methodology, introduced in Part III, is built around the same insight: the people who understand business rules and policy need to be able to verify that technical systems correctly implement them. COBOL's designers recognized this in 1959; the challenge has been maintaining that readability as systems grew in complexity over decades.

2. Bellotti, Marianne. *Kill It with Fire: Manage Aging Computer Systems (and Future Proof Modern Ones)*. No Starch Press, 2021. The opening of this chapter adapts and extends Bellotti's framing of the legacy system paradox from Chapter 1 of her book.

3. Bellotti, *Kill It with Fire*, Chapter 1, "Why Legacy Systems Are Worth Saving."

4. Martin Fowler, "Strangler Fig Application," martinfowler.com, August 22, 2024. https://martinfowler.com/bliki/StranglerFigApplication.html (accessed December 2025). Bellotti also discusses its application to legacy modernization in *Kill It with Fire*.

5. U.S. Government Accountability Office, "Information Technology: Agencies Need to Plan for Modernizing Critical Decades-Old Legacy Systems," GAO-25-107795 (July 2025), Table 1. https://www.gao.gov/products/gao-25-107795

6. GAO-25-107795, Appendix IV, System 3 profile. GAO substituted numeric identifiers for actual system names due to sensitivity concerns.

7. Even physical infrastructure faces technological evolution and changing requirements over time. Building materials improve; safety standards tighten; traffic patterns shift; accessibility requirements expand. A bridge built in 1980 might eventually need wider shoulders, different safety barriers, or structural reinforcement to meet current codes. The pace of change is simply much slower than in software, making the capital project model more workable—but not perfectly suited—even for physical assets.

8. GAO-25-107795, Figure 1, showing IT spending breakdown across 24 Chief Financial Officers Act agencies for fiscal year 2025.

9. When I worked in state government, I once had a legislator that was involved in budget negotiations push hard for focusing only on what were perceived to be "important" options for investing finite resource. He told me "if it's less than $100 million, I really don't want to hear about it right now."

10. GAO-25-107795, 13, citing U.S. Government Accountability Office, "Technology Modernization Fund: Although Planned Amounts Are Substantial, Projects Have Thus Far Achieved Minimal Savings," GAO-24-106575 (December 2023). https://www.gao.gov/products/gao-24-106575

11. Bellotti discusses the vendor lock-in dynamic in *Kill It with Fire*, noting how customization creates dependencies that make future changes more difficult.

12. GAO-25-107795, "What GAO Found" summary.

13. Melvin Conway, "How Do Committees Invent?" *Datamation*, April 1968.
http://www.melconway.com/Home/Committees_Paper.html
Bellotti discusses Conway's Law and its implications for legacy systems in *Kill It with Fire*, Chapter 2.

14. Bellotti, *Kill It with Fire*, Chapter 2, "Technical Debt Is Really Organizational Debt."

15. Partnership for Public Service, "A Profile of the 2023 Federal Workforce," analysis of OPM FedScope data.
https://ourpublicservice.org/fed-figures/workforce/ See also USAFacts, "Six charts on the age of federal workers," April 2025, analyzing OPM data from September 2024.
https://usafacts.org/articles/six-charts-on-the-age-of-federal-workers/

16. GAO-25-107795, Appendix IV, System 3 profile. For DOD's contract management system, the GAO found that "the average age of developers and technical subject matter experts on the system's team is above 60, putting the system at significant risk in being able to support and maintain it into the future."

17. California DMV modernization history: The first attempt (1987-1994) is documented in California State Auditor, "Department of Motor Vehicles: Its Automation Project Has Suffered Extensive Delays and Cost Overruns," Report 94-107 (April 1994). https://www.bsa.ca.gov/reports/summary/94107 The second attempt (2006-2013) is documented in California Legislative Analyst's Office, "DMV: Withdraw Funding for Database Project," (2013).
https://lao.ca.gov/Recommendations/Details/743

18. Bellotti, *Kill It with Fire*, Chapter 3, "The Rewrite Trap."

19. Brooks, Frederick P. *The Mythical Man-Month: Essays on Software Engineering*. Addison-Wesley, 1975. Chapter 5, "The Second-System Effect."

Chapter 3

1. Office of the Auditor General of Canada, "Building and Implementing the Phoenix Pay System" (May 2018). https://www.oag-bvg.gc.ca/internet/English/parl_oag_201805_01_e_43033.html By July 2018, Phoenix had caused pay problems for close to 80 percent of the federal government's 290,000 public servants through underpayments, overpayments, and non-payments.

2. Auditor General of Canada Michael Ferguson's characterization of the Phoenix project in his May 2018 report.

3. PSAC (Public Service Alliance of Canada), "Government downplays Phoenix pay system failures while workers continue to suffer nine years later" (February 2025). https://psacunion.ca/government-downplays-phoenix-pay-system-failures-while-workers As of January 2025, there were still 372,000 outstanding cases.

4. Michael Feathers, *Working Effectively with Legacy Code* (Prentice Hall, 2004). Feathers writes: "To me, legacy code is simply code without tests. Code without tests is bad code. It doesn't matter how well written it is; it doesn't matter how pretty or object-oriented or well-encapsulated it is."

5. Office of the Auditor General of Canada, "Report 1—Phoenix Pay Problems" (November 2017). https://www.oag-bvg.gc.ca/internet/English/parl_oag_201711_01_e_42666.html The report documents the complexity of federal pay rules and the extensive customization required.

6. Office of the Auditor General of Canada, "Building and Implementing the Phoenix Pay System" (May 2018). The report found that Phoenix executives "decided to defer or remove more than 100 important pay processing functions" and cancelled a pilot that would have assessed system readiness.

7. Martin Fowler, "Strangler Fig Application," martinfowler.com, August 22, 2024. https://martinfowler.com/bliki/StranglerFigApplication.html (accessed December 2025). The Strangler Fig pattern is named after vines that gradually envelop and replace their host trees, enabling systems to be modernized incrementally.

8. Office of the Auditor General of Canada, "Building and Implementing the Phoenix Pay System" (May 2018). https://www.oag-bvg.gc.ca/internet/English/parl_oag_201805_01_e_43033.html The report found that "deputy ministers from departments and agencies had no role in the Phoenix governance structure. They did not sit on any of the oversight bodies."

9. Feathers, *Working Effectively with Legacy Code*, Chapter 16: "I Don't Understand the Code Well Enough to Change It."

10. Office of the Auditor General of Canada, "Building and Implementing the Phoenix Pay System" (May 2018). https://www.oag-bvg.gc.ca/internet/English/parl_oag_201805_01_e_43033.html The report documents that Phoenix executives removed or deferred critical functionality including the ability to process retroactive pay requests.

11. Office of the Auditor General of Canada, "Report 1—Phoenix Pay Problems" (November 2017). https://www.oag-bvg.gc.ca/internet/English/parl_oag_201711_01_e_42666.html The report discusses how pay advisors at the Miramichi Pay Centre lacked complete and accurate procedures and could not deliver the level of service needed.

12. PSAC, "Government downplays Phoenix pay system failures while workers continue to suffer nine years later" (February 2025). https://psacunion.ca/government-downplays-phoenix-pay-system-failures-while-workers The report notes that "PSAC members who participated in recent demonstrations of Dayforce, the promised Phoenix replacement being tested in several departments, reported ongoing glitches and widespread issues."

Chapter 4

1. Mark Headd, "Hands Across America," Civic Innovations, March 2, 2021, https://civic.io/2021/03/02/hands-across-america/.

2. Superwhisper is an AI-powered voice-to-text app for Mac, Windows, and iOS that transcribes speech into text. https://superwhisper.com/

3. "GitHub Copilot," Wikipedia, https://en.wikipedia.org/wiki/GitHub_Copilot (accessed December 2025).

4. GitHub, "Under the hood: Exploring the AI models powering GitHub Copilot," *The GitHub Blog*, September 2, 2025. https://github.blog/ai-and-ml/github-copilot/under-the-hood-exploring-the-ai-models-powering-github-copilot/

5. "Claude (language model)," Wikipedia, https://en.wikipedia.org/wiki/Claude_(language_model) (accessed December 2025).

6. Anthropic, "Claude 3.7 Sonnet and Claude Code," February 24, 2025. https://www.anthropic.com/news/claude-3-7-sonnet

7. Anthropic, "Claude Opus 4.5," November 24, 2025. https://www.anthropic.com/news/claude-opus-4-5

8. Anthropic, "Claude Opus 4.5," November 24, 2025. https://www.anthropic.com/news/claude-opus-4-5

9. GitHub, "Copilot: Faster, smarter, and built for how you work now," *The GitHub Blog*, October 15, 2025. https://github.blog/news-insights/product-news/copilot-faster-smarter-and-built-for-how-you-work-now/

10. Microsoft Azure Blog, "How We Use AI Agents for COBOL Migration and Mainframe Modernization," July 16, 2025. https://azure.microsoft.com/en-us/blog/how-we-use-ai-agents-for-cobol-migration-and-mainframe-modernization/

11. Comment on "AI Tackles Aging COBOL Systems as Legacy Code Expertise Dwindles," *Slashdot*, April 24, 2025. https://developers.slashdot.org/story/25/04/24/1725256/ai-tackles-aging-cobol-systems-as-legacy-code-expertise-dwindles

12. GitHub, "How GitHub Copilot and AI agents are saving legacy systems," *The GitHub Blog*, October 14, 2025. https://github.blog/news-insights/research/how-github-copilot-and-ai-agents-are-saving-legacy-systems/

Chapter 5

1. Sean Grove, "The New Code," AI Engineer conference talk (2025). https://www.youtube.com/watch?v=8rABwKRsec4 Grove works on alignment research at OpenAI. The talk articulates specifications as the source artifact that should be preserved and version-controlled, with code as a derivative output that can be regenerated.

2. Grove, "The New Code."

3. Specifications are not, by themselves, a silver bullet for the challenges of managing context when using AI coding agents or persisting information across sessions. As discussed in Chapter 4, context window limitations remain real even as they have expanded dramatically—complex legacy systems often exceed even modern context limits, and information at the edges of the context window may be less reliably used. There is a developing ecosystem of strategies, techniques, and tools for managing context and agent "memory"—including retrieval-augmented generation, persistent knowledge bases, session summaries, project-level context files (such as AGENTS.md or .cursorrules), and emerging standards like Model Context Protocol (MCP) for connecting AI tools to external data sources. Specifications play an important role in this ecosystem but are one component of a broader approach to these challenges.

4. Grove, "The New Code."

5. Grove, "The New Code."

6. Mark Headd, "SpecOps: A New Approach to Legacy System Modernization," Civic Innovations, November 24, 2025. https://civic.io/2025/11/24/specops-a-new-approach-to-legacy-system-modernization/

7. SpecOps Instruction Sets, "Categories of Instruction Sets": Describes the three main categories of instruction sets used in SpecOps. https://github.com/spec-ops-method/spec-ops/blob/main/public/INSTRUCTION-SETS.md

8. SpecOps Instruction Sets, "Introduction": "A well-crafted instruction set for understanding COBOL code works whether

you're modernizing a benefits system in California or a tax system in New York."
https://github.com/spec-ops-method/spec-ops/blob/main/publi c/INSTRUCTION-SETS.md

9. SpecOps Instruction Sets, "The Evolving Nature of Instruction Sets": "When starting a SpecOps modernization project, you likely won't have all the instruction sets you need. This is normal and expected."
https://github.com/spec-ops-method/spec-ops/blob/main/publi c/INSTRUCTION-SETS.md

10. SpecOps Manifesto, "Specifications Are Technology-Agnostic": "A well-written specification describes what a system does without being tied to how it's implemented."
https://github.com/spec-ops-method/spec-ops/blob/main/publi c/MANIFESTO.md

11. Grove, "The New Code."

12. Grove, "The New Code."

13. Grove, "The New Code."

Chapter 6

1. The Weaveworks recovery story is recounted in multiple interviews with Alexis Richardson, including Software Engineering Radio Episode 440, "Alexis Richardson on GitOps" (December 2020), https://www.se-radio.net/2020/12/episode-440-alexis-richardso n-on-gitops/ and "How did GitOps get started?" interview with Schlomo Schapiro (February 2021), https://www.youtube.com/watch?v=lvLqJWOixDI Richardson describes how the team's disciplined approach to declarative infrastructure enabled rapid recovery from cluster failures.

2. "Infrastructure as Code: From Imperative to Declarative and Back Again," *The New Stack* (February 2025), traces the evolution from manual administration to declarative automation and the problems each phase addressed.
https://thenewstack.io/infrastructure-as-code-from-imperative-t o-declarative-and-back-again/

3. ArgoCD is a popular and widely used GitOps tool.https://argo-cd.readthedocs.io/en/stable/#what-is-argo-cd

4. Alexis Richardson, "Operations by Pull Request," Weaveworks blog, March 2017. https://web.archive.org/web/20230608145857/https://www.weave.works/blog/gitops-operations-by-pull-request The post described operational practices the Weaveworks team had developed while running Kubernetes in production.

5. Weaveworks, "What Is GitOps Really?" (May 2019). https://web.archive.org/web/20230531033515/https://www.weave.works/blog/what-is-gitops-really The four principles of GitOps have been adopted by the broader GitOps community. The principles emphasize declarative configuration, version control, automated deployment, and continuous reconciliation.

6. "Declarative vs. Imperative Programming for Infrastructure as Code (IaC)," Copado (November 2025), explains how declarative approaches handle configuration drift more effectively than imperative scripts because they describe desired end states rather than sequences of steps. https://www.copado.com/resources/blog/declarative-vs-imperative-infrastructure-as-code

7. SpecOps Manifesto, "Core Principles": "Like GitOps treats a Git repository as the single source of truth for cloud infrastructure, SpecOps treats the software specification as the authoritative description of system behavior." https://github.com/spec-ops-method/spec-ops/blob/main/public/MANIFESTO.md

8. Mark Headd, "SpecOps: A New Approach to Legacy System Modernization," Civic Innovations, November 24, 2025. https://civic.io/2025/11/24/specops-a-new-approach-to-legacy-system-modernization/

9. SpecOps Manifesto, "Changes Flow Through the Specification": "All system modifications begin with specification updates. This ensures changes are reviewed by appropriate stakeholders, validated against policy intent, and documented before implementation." https://github.com/spec-ops-method/spec-ops/blob/main/public/MANIFESTO.md

10. SpecOps Manifesto, "Specifications Should Be Accessible": "A specification should be readable by non-technical stakeholders while remaining detailed enough to guide implementation." https://github.com/spec-ops-method/spec-ops/blob/main/public/MANIFESTO.md

Chapter 7

1. The Policy2Code Prototyping Challenge ran from May through September 2024. https://beeckcenter.georgetown.edu/event/launch-information-session-policy2code-prototyping-challenge/

2. Ad Hoc Team, "Using generative AI to transform policy into code," *Ad Hoc Blog*, September 25, 2024. https://adhoc.team/2024/09/25/using-generative-ai-to-transform-policy-into-code/ The blog post describes the Policy2Code Prototyping Challenge approach: "Use Large Language Models (LLMs) to generate an intermediate format between policy language and software code that policy experts and software engineers could use to create a shared understanding."

3. SpecOps Introduction: "Traditional legacy modernization approaches—whether done through manual coding or AI-assisted conversion—focuses almost exclusively on one thing: converting old software code into new software code." https://github.com/spec-ops-method/spec-ops/blob/main/public/INTRO.md

4. SpecOps Comparison Framework: "Traditional AI-assisted modernization focuses on transpilation—the process of converting source code written in one programming language into equivalent source code in another programming language." https://github.com/spec-ops-method/spec-ops/blob/main/public/COMPARISON-FRAMEWORK.md

5. MITRE, "Legacy IT Modernization with AI," June 5, 2025. https://www.mitre.org/news-insights/publication/legacy-it-modernization-ai MITRE's IT Modernization team found that LLMs can reliably generate intermediate representations at scale from legacy code. Crucially, their research also showed that automated performance metrics do not match human subject-matter experts' perception of quality, meaning there is no substitute for human

verification. This empirically validates the SpecOps principle that domain experts must review AI-generated specifications.

6. Michael Feathers, *Working Effectively with Legacy Code* (Upper Saddle River, NJ: Prentice Hall, 2004). Feathers' definition of legacy code as "code without tests" and his techniques for characterization testing and finding seams provide the intellectual foundation for incremental legacy modernization approaches.

7. SpecOps Manifesto: "Traditional approaches to AI-assisted modernization treat the generated modern code as the primary output of value. SpecOps takes a fundamentally different view: the specification is what matters most." https://github.com/spec-ops-method/spec-ops/blob/main/public/MANIFESTO.md

8. Mark Headd, "What Does a Good Spec File Look Like?" Civic Innovations, December 2, 2025. https://civic.io/2025/12/02/what-does-a-good-spec-file-look-like/

9. The IRS Direct File project is available at https://github.com/IRS-Public/direct-file. The demonstration repository with skills files and generated specifications is available at https://github.com/spec-ops-method/spec-ops-demo.

10. Results from the SpecOps demonstration using IRS Direct File code samples. Evaluation criteria included accuracy of business logic extraction, clarity of specification language, and suitability for domain expert review.

11. U.S. Office of Personnel Management, "Retirement Statistics." https://www.opm.gov/retirement/statistics/ (accessed December 2025). The data shows 112,679 federal employees added to the Annuity Roll Processing System in FY 2025, continuing a pattern of approximately 100,000 annual retirements since FY 2000.

12. Natalie Alms, "Workforce cuts could complicate IRS goal to modernize in the next two years," *Nextgov/FCW*, June 12, 2025. https://www.nextgov.com/modernization/2025/06/workforce-cuts-could-complicate-irs-goal-modernize-next-two-years/401234

/ The article quotes a current IRS employee: "Systems break and it takes days to fix them because no one is left who knows how."

13. SpecOps Comparison Framework, "What Does Success Look Like?": "SpecOps has multiple success criteria, not just working code. Even if modernization is only partially complete, verified specifications for completed portions represent real value delivered."
https://github.com/spec-ops-method/spec-ops/blob/main/public/COMPARISON-FRAMEWORK.md

14. SpecOps Comparison Framework, "What Role Does AI Play?": "AI serves two distinct roles: 1. Understanding: Analyzing legacy code to extract and articulate system behavior. 2. Implementation: Generating modern code from verified specifications. Human expertise bridges these roles through specification review and verification."
https://github.com/spec-ops-method/spec-ops/blob/main/public/COMPARISON-FRAMEWORK.md

15. SpecOps Introduction: "Decoupling system behavior from technical implementation means governments never get stuck with decades-old legacy systems they desperately need to upgrade but don't fully understand. The specification preserves institutional knowledge regardless of the underlying technology."
https://github.com/spec-ops-method/spec-ops/blob/main/public/INTRO.md

16. SpecOps Introduction: "The name combines two proven approaches for building and managing modern software: Spec-driven development—focusing first on comprehensive descriptions of system behavior before code is written; and GitOps—treating version-controlled representations of cloud infrastructure as the ultimate source of truth for an environment."
https://github.com/spec-ops-method/spec-ops/blob/main/public/INTRO.md

17. SpecOps Manifesto: "The specification is what endures. The specification is what we can verify. The specification is what has value. Everything else is implementation details."
https://github.com/spec-ops-method/spec-ops/blob/main/public/MANIFESTO.md

Chapter 8

1. SpecOps Methodology, "Overview: The SpecOps Process."
https://github.com/spec-ops-method/spec-ops/blob/main/publi
c/METHODOLOGY.md

2. SpecOps FAQ, "Is SpecOps just waterfall disguised as agile?":
"Can you specify one component while implementing another?
Are specifications evolving based on learning? If yes, you're not
doing waterfall."
https://github.com/spec-ops-method/spec-ops/blob/main/FAQ.
md

3. SpecOps Methodology, Phase 1.3, "Knowledge Assessment."
https://github.com/spec-ops-method/spec-ops/blob/main/publi
c/METHODOLOGY.md

4. As discussed in Chapter 7, traditional modernizations do
produce some documents - requirements documents,
architecture diagrams, and test plans, etc. - but they almost
always exist solely to support the conversion process. Once that's
over, these documents are no longer actively used or maintained.

5. SpecOps Manifesto, "Specifications Enable Human
Verification": "Policy experts, program administrators, and
business stakeholders can't easily verify that a Java translation of
COBOL code is correct. But they can verify whether a
specification accurately describes eligibility rules, benefit
calculations, or tax logic."
https://github.com/spec-ops-method/spec-ops/blob/main/publi
c/MANIFESTO.md

6. Skills files and AI agent instruction sets are discussed more
fully in Chapter 5.

7. SpecOps Methodology, Phase 2.4, "Internal Technical Review."
https://github.com/spec-ops-method/spec-ops/blob/main/publi
c/METHODOLOGY.md

8. SpecOps Comparison Framework, "Who Verifies
Correctness?"
https://github.com/spec-ops-method/spec-ops/blob/main/publi
c/COMPARISON-FRAMEWORK.md

9. SpecOps FAQ, "What about AI hallucinations?": "Where hallucinations happen: Inventing business rules that don't exist, confidently stating incorrect logic, making up edge cases, fabricating policy references."
https://github.com/spec-ops-method/spec-ops/blob/main/FAQ.md

10. SpecOps Methodology, Phase 3.5, "Knowledge Capture."
https://github.com/spec-ops-method/spec-ops/blob/main/public/METHODOLOGY.md

11. SpecOps Manifesto, "Specifications Provide Direction to AI Agents": "A comprehensive specification serves as high-quality input for AI code generation. Rather than asking an AI agent to translate legacy code it may misunderstand, we're asking it to implement clearly articulated requirements."
https://github.com/spec-ops-method/spec-ops/blob/main/public/MANIFESTO.md

12. SpecOps Methodology, Phase 5.1, "Specification Conformance Testing."
https://github.com/spec-ops-method/spec-ops/blob/main/public/METHODOLOGY.md

13. SpecOps Methodology, Phase 6.1, deployment strategies including Strangler Fig.
https://github.com/spec-ops-method/spec-ops/blob/main/public/METHODOLOGY.md

14. SpecOps Manifesto, "Specifications Are Technology-Agnostic": "A well-written specification describes what a system does without being tied to how it's implemented. This means the same specification can guide multiple implementations."
https://github.com/spec-ops-method/spec-ops/blob/main/public/MANIFESTO.md

15. SpecOps Methodology, "Working Incrementally with the Strangler Fig Pattern."
https://github.com/spec-ops-method/spec-ops/blob/main/public/METHODOLOGY.md

16. Martin Fowler, "Strangler Fig Application," martinfowler.com, August 22, 2024.

https://martinfowler.com/bliki/StranglerFigApplication.html
(accessed December 2025). The Strangler Fig pattern is named
after vines that gradually envelop and replace their host trees,
enabling systems to be modernized incrementally.

17. SpecOps Methodology, "SpecOps Strangler Fig
Implementation Strategy."
https://github.com/spec-ops-method/spec-ops/blob/main/publi
c/METHODOLOGY.md

Chapter 9

1. Mark Headd, "Win big by going small," *18F Blog*, March
13, 2018.
https://blog.18f.org/2018/03/13/win-big-by-going-small/

2. SpecOps Manifesto, "Core Principles": "Like GitOps treats
a Git repository as the single source of truth for cloud
infrastructure, SpecOps treats the software specification as the
authoritative description of system behavior."
https://github.com/spec-ops-method/spec-ops/blob/main/publi
c/MANIFESTO.md

3. SpecOps Manifesto, "Why Specifications Are the Valuable
Artifact": "A well-written specification describes what a system
does without being tied to how it's implemented. This means the
same specification can guide multiple implementations—different
technology stacks, different deployment models, even different
functional decompositions."
https://github.com/spec-ops-method/spec-ops/blob/main/publi
c/MANIFESTO.md

4. As outlined in Chapter 8, teams don't create a
specification for an entire system, then verify it all, then
implement it all. They work component by component: select a
component, move it through all six phases of SpecOps, deploy it,
then select the next component and repeat. The process is
iterative.

5. U.S. Census Bureau, "By 2030, All Baby Boomers Will Be
Age 65 or Older," December 10, 2019.
https://www.census.gov/library/stories/2019/12/by-2030-all-ba
by-boomers-will-be-age-65-or-older.html

6. SpecOps Manifesto, "Specifications Capture Institutional Knowledge": "When we generate a specification, we're not just documenting syntax—we're recovering and preserving institutional knowledge before it's lost forever."
https://github.com/spec-ops-method/spec-ops/blob/main/public/MANIFESTO.md

7. Marianne Bellotti, *Kill It with Fire: Manage Aging Computer Systems (and Future Proof Modern Ones)* (San Francisco: No Starch Press, 2021). Bellotti's experience at the U.S. Digital Service informs her analysis of legacy systems and why they deserve respect before replacement.

8. SpecOps Comparison Framework, "Verification and Validation": Discusses the verification gap in traditional modernization approaches.
https://github.com/spec-ops-method/spec-ops/blob/main/public/COMPARISON-FRAMEWORK.md

9. SpecOps Manifesto, "Specifications Enable Human Verification": "Policy experts, program administrators, and business stakeholders can't easily verify that a Java translation of COBOL code is correct. But they can verify whether a specification accurately describes eligibility rules, benefit calculations, or tax logic."
https://github.com/spec-ops-method/spec-ops/blob/main/public/MANIFESTO.md

10. SpecOps Manifesto, "AI Agents Are Tools for Both Understanding and Implementation": "AI excels at two distinct tasks in modernization: analyzing legacy code to extract behavior patterns, and generating new code from clear specifications."
https://github.com/spec-ops-method/spec-ops/blob/main/public/MANIFESTO.md

11. SpecOps Manifesto, "Changes Flow Through the Specification": "All system modifications begin with specification updates. This ensures changes are reviewed by appropriate stakeholders, validated against policy intent, and documented before implementation."
https://github.com/spec-ops-method/spec-ops/blob/main/public/MANIFESTO.md

12. SpecOps Manifesto, "Specifications Should Be Accessible": "A specification should be readable by non-technical stakeholders while remaining detailed enough to guide implementation."
https://github.com/spec-ops-method/spec-ops/blob/main/public/MANIFESTO.md

13. SpecOps Core Tools outlines the four main categories of tools required for SpecOps implementation.
https://github.com/spec-ops-method/spec-ops/blob/main/public/CORE-TOOLS.md

14. SpecOps FAQ, "What tools do I need to get started?": "Don't wait for perfect tooling. Start with what you have."
https://github.com/spec-ops-method/spec-ops/blob/main/FAQ.md

15. SpecOps Instruction Sets, "Introduction": "A key insight: Instruction sets are more portable and sharable than code. A well-crafted instruction set for understanding COBOL code works whether you're modernizing a benefits system in California or a tax system in New York."
https://github.com/spec-ops-method/spec-ops/blob/main/public/INSTRUCTION-SETS.md

16. SpecOps Manifesto, "Specifications Are Technology-Agnostic": "A well-written specification describes what a system does without being tied to how it's implemented."
https://github.com/spec-ops-method/spec-ops/blob/main/public/MANIFESTO.md

17. Bellotti, *Kill It with Fire*. Her analysis emphasizes that modernization failures often stem from organizational dynamics rather than purely technical limitations.

18. Bellotti, *Kill It with Fire*. Her emphasis on understanding before acting and incremental replacement aligns closely with SpecOps principles.

Chapter 10

1. Partnership for Public Service, "A Profile of the 2023 Federal Workforce," analysis of OPM FedScope data.

https://ourpublicservice.org/fed-figures/workforce/ The average federal worker age of 47.2 compares to Bureau of Labor Statistics data showing median age of 42.2 for the U.S. labor force in 2024.

2. USAFacts, "Six charts on the age of federal workers," April 2025, analyzing OPM data from September 2024. https://usafacts.org/articles/six-charts-on-the-age-of-federal-workers/ The 28 percent figure represents full-time permanent employees age 55 or above.

3. Partnership for Public Service, "A Profile of the 2023 Federal Workforce." https://ourpublicservice.org/fed-figures/workforce/ The disparity in under-30 representation (7 percent federal versus nearly 20 percent private sector) has persisted for over a decade.

4. USAFacts analysis of OPM FedScope data, tracking age distribution changes from 2005 to 2024. https://usafacts.org/articles/six-charts-on-the-age-of-federal-workers/

5. Tom Taulli, "COBOL Language: Call It A Comeback," *Forbes*, July 13, 2020. https://www.forbes.com/sites/tomtaulli/2020/07/13/cobol-language-call-it-a-comeback/ See also Chapter 1 discussion of state unemployment systems during the COVID-19 pandemic.

6. OPM Director Scott Kupor, statement on federal workforce changes, November 2025. The 317,000 figure includes retirements, voluntary separations, early retirement incentives, and involuntary separations.

7. Ibid. The 68,000 new hires were concentrated in priority areas including immigration and border enforcement.

8. U.S. Office of Personnel Management, "Retirement Statistics." https://www.opm.gov/retirement/statistics/ (accessed December 2025). The annuity roll data shows 112,679 federal employees added to retirement rolls in FY 2025, consistent with annual totals ranging from 75,000 to 115,000 since FY 2000.

9. U.S. Government Accountability Office, "Information Technology: Agencies Need to Plan for Modernizing Critical Decades-Old Legacy Systems," GAO-25-107795 (July 2025).

https://www.gao.gov/products/gao-25-107795 The DOD contract management system was initially developed in 1964 and still runs on COBOL and assembly language code.

10. Natalie Alms, "Workforce cuts could complicate IRS goal to modernize in the next two years," *Nextgov/FCW*, June 12, 2025. https://www.nextgov.com/modernization/2025/06/workforce-cuts-could-complicate-irs-goal-modernize-next-two-years/401234 / The article reports significant departures from the IRS Chief Information Officer organization.

11. Ibid. The quotation is from a current IRS employee describing operational impacts of workforce departures.

Chapter 11

1. At the time of the rollout, the state's Office of Information Technology (OIT) was under increased scrutiny from some recently failed technology implementations. In the early years of the next administration, the beleaguered OIT would be replaced by a new technology department called the Department of Technology and Information (DTI). https://archives.delaware.gov/delaware-agency-histories/department-technology-information-dti/

2. Governor Ruth Ann Minner, State of the State Address, Delaware FY 2003 Operating Budget, January 2002. https://budget.delaware.gov/budget/fy2003/state-of-state.shtml

3. Delaware Department of Technology and Information, "Current IT Projects: ERP Modernization." The project description notes that PHRST is among "three 13 to 20+ year old statewide PeopleSoft applications" being modernized. https://dti.delaware.gov/digital-innovation/current-it-projects/

4. Office of the Auditor General of Canada, "Building and Implementing the Phoenix Pay System" (May 2018). https://www.oag-bvg.gc.ca/internet/English/parl_oag_201805_01_e_43033.html The system has affected over 80 percent of federal employees and costs continue to accumulate for remediation efforts.

5. Brianna Hill, "Delaware moves to correct decades-old legislative pension error," Spotlight Delaware, January 15, 2025. The article details how a 1997 Compensation Commission recommendation was never properly codified into law due to a clerical oversight, resulting in $900,000 in retroactive payments when the error was discovered in 2024. https://spotlightdelaware.org/2025/01/15/delaware-legislative-pension-fix/

6. Ibid.

Chapter 12

1. The $12 billion estimate comes from sponsors of the SHARE IT Act. See H.R. 9566, Source Code Harmonization And Reuse in Information Technology Act. https://www.congress.gov/bill/118th-congress/house-bill/9566 The broader context: the federal government spends over $100 billion annually on IT, with approximately 80% going to operations and maintenance of existing systems.

2. Office of Management and Budget, M-16-21, "Federal Source Code Policy: Achieving Efficiency, Transparency, and Innovation through Reusable and Open Source Software" (August 8, 2016). https://obamawhitehouse.archives.gov/sites/default/files/omb/memoranda/2016/m_16_21.pdf The policy requires agencies to conduct a three-step analysis for software needs: conduct strategic analysis, consider existing commercial solutions, then consider custom development.

3. The Source Code Harmonization And Reuse in Information Technology Act (H.R. 9566) was signed into law by President Biden on December 23, 2024. https://www.congress.gov/bill/118th-congress/house-bill/9566 The law requires agency CIOs to develop policies within 180 days ensuring custom-developed code aligns with best practices and establishing processes for making code metadata publicly available.

4. Login.gov is a shared service provided by the General Services Administration's Technology Transformation Services. https://login.gov/ It provides secure sign-in services for participating government agencies, handling identity verification

and authentication so individual agencies don't have to build and maintain their own systems.

5. Cloud.gov is a Platform-as-a-Service built specifically for government, also provided by GSA's Technology Transformation Services. https://cloud.gov/ It handles FedRAMP compliance at the platform level, reducing the compliance burden on individual agency applications.

6. The U.S. Web Design System (USWDS) provides design principles, components, and code patterns for building accessible, mobile-friendly government websites. https://designsystem.digital.gov/ It's maintained by GSA's Technology Transformation Services and used by hundreds of federal websites.

7. SpecOps Instruction Sets, "Sharing Instruction Sets Across Governments": "Many governments use the same legacy technologies. COBOL is COBOL everywhere. Mainframe patterns are similar across organizations. Database structures follow common patterns. Business logic differs, but technical patterns don't." https://github.com/spec-ops-method/spec-ops/blob/main/public/INSTRUCTION-SETS.md

8. SpecOps FAQ, "What can be shared vs. what must stay internal?": "If it's knowledge about platforms, languages, or general patterns—share it. If it's specific to your agency's programs or systems—keep it internal." https://github.com/spec-ops-method/spec-ops/blob/main/FAQ.md

9. SpecOps Instruction Sets, "Conclusion: Building Together": "By building a shared library of instruction sets, the government technology community can reduce duplicated effort across agencies, accelerate adoption of SpecOps methodology, preserve knowledge about legacy systems, lower barriers to modernization, and create a lasting resource for future projects." https://github.com/spec-ops-method/spec-ops/blob/main/public/INSTRUCTION-SETS.md

10. SpecOps Instruction Sets, "Instruction Set Repository Structure": Proposes organizing shared instruction sets by category including legacy-languages, legacy-platforms, domains,

specification, implementation, government-specific, and templates. https://github.com/spec-ops-method/spec-ops/blob/main/public/INSTRUCTION-SETS.md

Chapter 13

1. California DMV modernization history: The first attempt (1987-1994) is documented in California State Auditor, "Department of Motor Vehicles: Its Automation Project Has Suffered Extensive Delays and Cost Overruns," Report 94-107 (April 1994). https://www.bsa.ca.gov/reports/summary/94107 The second attempt (2006-2013) is documented in California Legislative Analyst's Office, "DMV: Withdraw Funding for Database Project," (2013). https://lao.ca.gov/Recommendations/Details/743

2. SpecOps Methodology, Phase 1.6, "Pilot Selection": "Select a moderate-complexity component with known behavior... Choose something small enough to complete in 2-4 weeks." https://github.com/spec-ops-method/spec-ops/blob/main/public/METHODOLOGY.md

3. SpecOps FAQ, "How big does my team need to be?": "Minimum viable team: 8-10 people." https://github.com/spec-ops-method/spec-ops/blob/main/FAQ.md

4. SpecOps Team Structure, "Business Analyst / Technical Writer" role description. https://github.com/spec-ops-method/spec-ops/blob/main/public/TEAM-STRUCTURE.md

5. SpecOps Team Structure, "Core Team Structure": "Total Core Team Size: 12-18 people (with varying levels of time commitment)." https://github.com/spec-ops-method/spec-ops/blob/main/public/TEAM-STRUCTURE.md

6. SpecOps FAQ, "Do I need people who know the legacy language?": "The shift: From 'need rare COBOL experts' to 'need domain experts + AI-skilled engineers.'"

https://github.com/spec-ops-method/spec-ops/blob/main/FAQ.
md

7. SpecOps FAQ, "Can contractors do this or does it require government staff?": "Hybrid approach works best."
https://github.com/spec-ops-method/spec-ops/blob/main/FAQ.
md

8. SpecOps Team Structure, "Red Flags to Avoid."
https://github.com/spec-ops-method/spec-ops/blob/main/publi
c/TEAM-STRUCTURE.md

9. SpecOps FAQ, "What tools do I need to get started?": Lists the minimum viable toolchain components.
https://github.com/spec-ops-method/spec-ops/blob/main/FAQ.
md

10. GitHub, "Under the hood: Exploring the AI models powering GitHub Copilot," *The GitHub Blog*, September 2, 2025.
https://github.blog/ai-and-ml/github-copilot/under-the-hood-ex
ploring-the-ai-models-powering-github-copilot/

11. SpecOps Instruction Sets, "Introduction": "A key insight: Instruction sets are more portable and sharable than code."
https://github.com/spec-ops-method/spec-ops/blob/main/publi
c/INSTRUCTION-SETS.md

12. SpecOps Instruction Sets, "Instruction Set Maturity Levels."
https://github.com/spec-ops-method/spec-ops/blob/main/publi
c/INSTRUCTION-SETS.md

13. SpecOps FAQ, "How long does it take to see results?": Provides the month-by-month timeline for a pilot component.
https://github.com/spec-ops-method/spec-ops/blob/main/FAQ.
md

14. SpecOps FAQ, "How long does it take to see results?": "Compare to: Traditional manual approach: 6-12 months for same component. Direct AI translation: 1-2 months but higher risk of errors."
https://github.com/spec-ops-method/spec-ops/blob/main/FAQ.
md

15. SpecOps FAQ, "How do I get executive buy-in?": Provides messaging framework for leadership conversations.

https://github.com/spec-ops-method/spec-ops/blob/main/FAQ.md

Chapter 14

1. USDS Digital Services Playbook, Play 7: "Bring in experienced teams." https://playbook.cio.gov/#play7 The playbook emphasizes that teams should include people with experience building digital services and subject matter expertise in the relevant policy domain.

2. USDS Digital Services Playbook, Play 4: "Build the service using agile and iterative practices." https://playbook.cio.gov/#play4 The playbook calls for shipping a functioning minimum viable product quickly, running usability tests frequently, and releasing features and improvements multiple times each month.

3. 18F Partnership Principles, "An empowered product owner": "...an empowered product owner from your agency who understands your organization, the problem we're solving, and can advocate for the product we ultimately build together." https://web.archive.org/web/20241203184424/https://18f.gsa.gov/partnership-principles/

4. SpecOps Team Structure, "Training Needs for the Team": Domain experts need training on "how to read and evaluate technical specifications" and the "SpecOps verification process." https://github.com/spec-ops-method/spec-ops/blob/main/public/TEAM-STRUCTURE.md

5. SpecOps Team Structure, "Red Flags to Avoid": "Technical staff who dismiss domain expert input" is identified as a warning sign that undermines the methodology. https://github.com/spec-ops-method/spec-ops/blob/main/public/TEAM-STRUCTURE.md

6. SpecOps Instruction Sets, "Instruction Set Maturity Levels": Describes how instruction sets mature through use from "Level 1: Initial" rough drafts to "Level 2: Developing" and eventually "Level 3: Established" as teams refine them based on results. https://github.com/spec-ops-method/spec-ops/blob/main/public/INSTRUCTION-SETS.md

7. SpecOps FAQ, "What are the biggest risks?": Lists "AI generates poor specifications" as a top risk, with the mitigation being "technical review before domain review, multiple verification methods, explicit uncertainty documentation." https://github.com/spec-ops-method/spec-ops/blob/main/FAQ.md

8. Gene Kim and Steve Yegge, *Vibe Coding: Building Production-Grade Software With GenAI, Chat, Agents, and Beyond* (IT Revolution, 2025), Chapter 18 on building organizational culture for AI-assisted development.

Chapter 15

1. SpecOps Methodology, Phase 1.4, "Component Prioritization," lists "knowledge loss risk" as a prioritization criterion alongside business criticality and complexity. https://github.com/spec-ops-method/spec-ops/blob/main/public/METHODOLOGY.md

2. SpecOps Instruction Sets describes how "domain experts outside the team" and "legacy system specialists" including "consultants or retired experts" can contribute to instruction set development without being full-time team members. https://github.com/spec-ops-method/spec-ops/blob/main/public/INSTRUCTION-SETS.md

3. SpecOps FAQ, "What are the biggest risks?": Lists strategies for managing uncertainty including "explicit uncertainty documentation" and "multiple verification methods." https://github.com/spec-ops-method/spec-ops/blob/main/FAQ.md

4. SpecOps Instruction Sets, "Sharing Instruction Sets Across Governments": "Common Legacy Platforms: Many governments use the same legacy technologies. COBOL is COBOL everywhere. Mainframe patterns are similar across organizations." https://github.com/spec-ops-method/spec-ops/blob/main/public/INSTRUCTION-SETS.md

5. SpecOps FAQ, "What about AI hallucinations?": "Where hallucinations happen: Inventing business rules that don't exist, confidently stating incorrect logic, making up edge cases,

fabricating policy references."
https://github.com/spec-ops-method/spec-ops/blob/main/FAQ.md

6. SpecOps Methodology, Phase 1.3, "Knowledge Assessment," includes interviewing stakeholders about system behavior and documenting "critical business rules and edge cases people remember."
https://github.com/spec-ops-method/spec-ops/blob/main/public/METHODOLOGY.md

7. SpecOps FAQ, "What if the legacy system has bugs we want to fix?": "During verification: Domain experts identify bugs. Decide: fix in modern implementation or preserve bug (compatibility). Document decision rationale."
https://github.com/spec-ops-method/spec-ops/blob/main/FAQ.md

8. SpecOps FAQ provides a template for documenting bug decisions including "Legacy behavior," "Policy intent," "Decision for modern implementation," and "Verification" of the decision by appropriate authorities.
https://github.com/spec-ops-method/spec-ops/blob/main/FAQ.md

9. As discussed in Chapter 8, verification catches errors that would otherwise propagate into the new system: "Every error not caught in verification becomes an error in implementation, discovered later when it's far more expensive to fix."

10. SpecOps FAQ, "How long does SpecOps take?": Notes that timeline depends on factors including "team experience with AI tools" and "quality of existing documentation," with experienced teams working faster.
https://github.com/spec-ops-method/spec-ops/blob/main/FAQ.md

Chapter 16

1. Office of the Auditor General of Canada, "Building and Implementing the Phoenix Pay System" (May 2018).
https://www.oag-bvg.gc.ca/internet/English/parl_oag_201805_01_e_43033.html The system continues to generate problems

nearly a decade after launch, with over 370,000 outstanding pay disputes as of 2025.

2. This insight is developed across several chapters. Chapter 5 introduces Sean Grove's framing of specifications as source code: "The prompt is the source. The code is the compiled output." Chapter 7 distinguishes compilation (preserving knowledge in specifications) from transpilation (translating code without preserving understanding). Chapter 9 establishes "the specification is the source of truth" as the first core principle of SpecOps.

3. SpecOps Manifesto, "Changes Flow Through the Specification": "All system modifications begin with specification updates. This ensures changes are reviewed by appropriate stakeholders, validated against policy intent, and documented before implementation."
https://github.com/spec-ops-method/spec-ops/blob/main/publi c/MANIFESTO.md

4. Martin Fowler, "StranglerFigApplication," martinfowler.com, June 29, 2004.
https://martinfowler.com/bliki/StranglerFigApplication.html (accessed December 2025). The Strangler Fig pattern is named after vines that gradually envelop and replace their host trees, enabling systems to be modernized incrementally. See also Chapter 8's treatment of incremental modernization.

5. Chapter 12 examines why direct code sharing has disappointed expectations while instruction sets offer a more promising collaboration model. The key insight: "COBOL is COBOL everywhere. Mainframe patterns are similar across organizations. Database structures follow common patterns. Business logic differs, but technical patterns don't."

6. Chapter 11 discusses accountability requirements unique to government and how specifications serve oversight needs that code can't: "Code embeds business logic in implementation details that auditors can't read. When systems produce unexpected results, explaining why requires technical expertise that oversight bodies don't have."

7. Chapter 2, "The Budget Structure That Prevents Continuous Improvement," traces how budget processes designed for physical

infrastructure create a project-based funding model unsuited to software that requires continuous development.

8. Chapter 4 discusses how AI coding assistants change the skill mix required for legacy modernization: "The shift is from 'need rare COBOL experts' to 'need domain experts + AI-skilled engineers.' Domain knowledge becomes more valuable relative to legacy language expertise."

9. Chapter 14 describes how SpecOps positions domain experts: "In most legacy modernization projects, domain experts participate at the edges... SpecOps takes a different approach. Domain expert verification sits at the center of the methodology, not the periphery."

10. Chapter 1 describes this convergence: "The workforce crisis is reaching an acute phase. Retirements are accelerating. The pandemic demonstrated how fragile these systems are under stress. At the same time, AI capabilities have reached a threshold of usefulness for exactly the kind of knowledge extraction and translation that legacy modernization requires."

11. Chapter 10 provides detailed workforce demographics. Federal workforce data from USAFacts analysis of OPM FedScope data and Partnership for Public Service analysis. COBOL programmer age data from industry sources including AFCEA's *SIGNAL Magazine*.

Chapter 17

1. "Test-driven development," Wikipedia, https://en.wikipedia.org/wiki/Test-driven_development (accessed December 2025).

2. "Behavior-driven development," Wikipedia, https://en.wikipedia.org/wiki/Behavior-driven_development (accessed December 2025).

About the Author

Mark Headd is the former Chief Data Officer for the City of Philadelphia, serving as one of the first municipal Chief Data Officers in the United States. Self-taught in software development, he holds a Master's degree in Public Administration from the Maxwell School of Citizenship and Public Affairs at Syracuse University, and is a former adjunct instructor at the University of Delaware's Institute for Public Administration.

Mark is also the author of *How to Talk to Civic Hackers* (2016), an open-source book on government collaboration with technology communities, available under a Creative Commons license at https://www.civichacking.guide/

He lives in and works from Syracuse, New York.

www.ingramcontent.com/pod-product-compliance
Lightning Source LLC
Chambersburg PA
CBHW071534200326
41519CB00021BB/6481